CONTENTS

THE ROLE OF STRATEGIC AND CRITICAL MINERALS IN OUR NATIONAL AND ECONOMIC SECURITY

OVERSIGHT HEARING

BEFORE THE

SUBCOMMITTEE ON ENERGY AND MINERAL RESOURCES

OF THE

COMMITTEE ON RESOURCES
U.S. HOUSE OF REPRESENTATIVES

ONE HUNDRED EIGHTH CONGRESS

FIRST SESSION

Thursday, July 17, 2003

Serial No. 108-42

Printed for the use of the Committee on Resources

Available via the World Wide Web: http://www.access.gpo.gov/congress/house
or
Committee address: http://resourcescommittee.house.gov

U.S. GOVERNMENT PRINTING OFFICE

88-357 PS WASHINGTON : 2003

For sale by the Superintendent of Documents, U.S. Government Printing Office
Internet: bookstore.gpo.gov Phone: toll free (866) 512–1800; DC area (202) 512–1800
Fax: (202) 512–2250 Mail: Stop SSOP, Washington, DC 20402–0001

COMMITTEE ON RESOURCES

RICHARD W. POMBO, California, *Chairman*
NICK J. RAHALL II, West Virginia, *Ranking Democrat Member*

Don Young, Alaska
W.J. "Billy" Tauzin, Louisiana
Jim Saxton, New Jersey
Elton Gallegly, California
John J. Duncan, Jr., Tennessee
Wayne T. Gilchrest, Maryland
Ken Calvert, California
Scott McInnis, Colorado
Barbara Cubin, Wyoming
George Radanovich, California
Walter B. Jones, Jr., North Carolina
Chris Cannon, Utah
John E. Peterson, Pennsylvania
Jim Gibbons, Nevada,
Vice Chairman
Mark E. Souder, Indiana
Greg Walden, Oregon
Thomas G. Tancredo, Colorado
J.D. Hayworth, Arizona
Tom Osborne, Nebraska
Jeff Flake, Arizona
Dennis R. Rehberg, Montana
Rick Renzi, Arizona
Tom Cole, Oklahoma
Stevan Pearce, New Mexico
Rob Bishop, Utah
Devin Nunes, California
Randy Neugebauer, Texas

Dale E. Kildee, Michigan
Eni F.H. Faleomavaega, American Samoa
Neil Abercrombie, Hawaii
Solomon P. Ortiz, Texas
Frank Pallone, Jr., New Jersey
Calvin M. Dooley, California
Donna M. Christensen, Virgin Islands
Ron Kind, Wisconsin
Jay Inslee, Washington
Grace F. Napolitano, California
Tom Udall, New Mexico
Mark Udall, Colorado
Aníbal Acevedo-Vilá, Puerto Rico
Brad Carson, Oklahoma
Raúl M. Grijalva, Arizona
Dennis A. Cardoza, California
Madeleine Z. Bordallo, Guam
George Miller, California
Edward J. Markey, Massachusetts
Rubén Hinojosa, Texas
Ciro D. Rodriguez, Texas
Joe Baca, California
Betty McCollum, Minnesota

Steven J. Ding, *Chief of Staff*
Lisa Pittman, *Chief Counsel*
James H. Zoia, *Democrat Staff Director*
Jeffrey P. Petrich, *Democrat Chief Counsel*

SUBCOMMITTEE ON ENERGY AND MINERAL RESOURCES

BARBARA CUBIN, Wyoming, *Chairman*
RON KIND, Wisconsin, *Ranking Democrat Member*

W.J. "Billy" Tauzin, Louisiana
Chris Cannon, Utah
Jim Gibbons, Nevada
Mark E. Souder, Indiana
Dennis R. Rehberg, Montana
Tom Cole, Oklahoma
Stevan Pearce, New Mexico
Rob Bishop, Utah
Devin Nunes, California
Richard W. Pombo, California, *ex officio*

Eni F.H. Faleomavaega, American Samoa
Solomon P. Ortiz, Texas
Grace F. Napolitano, California
Tom Udall, New Mexico
Brad Carson, Oklahoma
Edward J. Markey, Massachusetts
VACANCY
Nick J. Rahall II, West Virginia, *ex officio*

OVERSIGHT HEARING ON "THE ROLE OF STRATEGIC AND CRITICAL MINERALS IN OUR NATIONAL AND ECONOMIC SECURITY."

Thursday, July 17, 2003
U.S. House of Representatives
Subcommittee on Energy and Mineral Resources
Committee on Resources
Washington, DC

The Subcommittee met, pursuant to call, at 10:11 a.m., in room 1334, Longworth House Office Building, Hon. Barbara Cubin, Chairman of the Subcommittee, presiding.

Present: Representatives Cubin, Gibbons, Cole, Nunes, Kind, Napolitano, and Tom Udall.

Mrs. CUBIN. The oversight hearing by the Subcommittee on Energy and Mineral Resources will please come to order.

The Subcommittee is meeting today to hear testimony on the role of strategic and critical minerals in our national and economic security. Under Rule 4(g) the Chairman and the Ranking Minority Member can make opening statements. Any other members who have statements can have their statements included in the record under unanimous consent. Excuse me, they can always have their statement included in the record, but if you want to say something, Mr. Gibbons, then you will have to have unanimous consent for that.

Mr. GIBBONS. Feeling generous.

Mrs. CUBIN. Yes, that is right, and I am feeling tough.

[Laughter.]

Mr. GIBBONS. I noticed you picked on me.

Mrs. CUBIN. Well, you were the only one over there.

[Laughter.]

Mrs. CUBIN. Mr. Kind is over here.

STATEMENT OF THE HON. BARBARA CUBIN. A REPRESENTATIVE IN CONGRESS FROM THE STATE OF WYOMING

Mrs. CUBIN. The Subcommittee meets today to discuss a subject with both economic and national security implications. Our Nation is facing a dwindling production of strategic and critical minerals and a growing reliance on foreign imports to supply minerals to our defense, aerospace and high-tech industries. America is blessed with an abundant supply of metals and other minerals. That mineral supply has played a critical role in America's economic development, our success in technological development and our national security.

Our domestic mineral supply comes from our domestic hard rock mining industry. Historically the U.S. has been a world leader in the mining of a variety of minerals, and continues to be one of the largest producers of copper, gold, lead, zinc and silver. We have substantial reserves of these and many other important minerals. Minerals are key to any manufacturing economy and used extensively in modern industries such as aerospace, communications, robotics and information processing. If our Nation is going to be the leader in technology, it must have access to critical minerals, preferably a secure supply produced domestically. Minerals have a profound effect on our Nation's economic security.

Minerals availability is also crucial to America's ability to defend itself, maintain its position as a global superpower and carry out the war on terror. As several of our witnesses will testify today, the United States is becoming increasingly dependent on foreign sources for strategic minerals that are used by our defense, security and aerospace industries. Nations such as China, Russia, Morocco and South Africa are now providing minerals that are key to our national security. This is a disturbing situation.

Our Nation is becoming ever more dependent on minerals imports because our current policies discourage domestic production. There are some alarming trends occurring in our domestic mining industry. The United States' share of worldwide hard rock minerals exploration dropped 10 percent in 2000 alone. Also, since the mid 1990's there has been a sharp decline in mining claims. Further, since 1997 there has been a 66 percent increase in exploration spending in the United States. These companies are investing abroad. I said increase. A 66 percent decrease in explorations spending in the United States. Pardon me. These companies are investing abroad.

Why are we losing our domestic industry? First, political risk for mining companies is higher in the United States than in almost all other countries, including those nations many in this room would not travel to themselves. Second, regulatory changes in policies that have been initiated over the past decade are destroying our domestic mining industry. The Millsite opinion issued in 1997 and the rewrite of the 3809 regulations in 2001 created massive uncertainty for anyone developing a mine plan in the United States. Third, the land withdrawals, such as the California Desert Protection Act and the Roadless Rule, have jeopardized exploration and development of mineral resources adjacent to withdrawn areas and has locked up millions of acres with promising mineral potential. Uncertainty in the permitting process and frivolous litigation are delaying projects for up to 10 years. Finally, an extended crisis in the surety bond industry has made it virtually impossible for mining companies to obtain reclamation insurance bonds, thus making it exceedingly difficult to obtain permits. All of these factors are destroying our domestic mining industry. Once again, our Federal lands policies are running counter to America's economic health and national security.

If we want to remain a strong and vibrant economic and military force, we need to evaluate our current minerals policies and develop one based on common sense. We need policies that promote American industries and American jobs.

I thank you all for coming today and look forward to the testimony of our witnesses.

[The prepared statement of Mrs. Cubin follows:]

Statement of The Honorable Barbara Cubin, Chairman, Subcommittee on Energy and Mineral Resources

The Subcommittee meets today to discuss a subject with both economic and national security implications. Our Nation is facing a dwindling production of strategic and critical minerals and a growing reliance on foreign imports to supply minerals to our defense, aerospace and high-tech industries.

America is blessed with an abundant supply of metals and other minerals. That mineral supply has played a crucial role in America's economic development, our success in technological development and our national security.

Our domestic mineral supply comes from our domestic hardrock mining industry. Historically, the U.S. has been a world leader in the mining of a variety of minerals, and continues to be one of the largest producers of copper, gold, lead, silver and zinc.

We have substantial reserves of these and many other important minerals. Minerals are key to any manufacturing economy and are used extensively in modern industries such as aerospace, communications, robotics and information processing.

If our nation is going to be the leader in technology, it must have access to critical minerals, preferably a secure supply produced domestically. Minerals have a profound affect on our nation's economic security.

Minerals availability is also crucial to America's ability to defend itself, maintain its position as global superpower and carry out the War on Terror. As several of our witnesses will testify today, the U.S. is becoming increasingly dependent on foreign sources for strategic minerals that are used by our defense, security and aerospace industries.

Nations such as China, Russia, Morocco, Gabon and South Africa are now providing minerals that are key to our national security. This is a disturbing situation.

Our nation is becoming ever more dependent on minerals imports because our current policies discourage domestic production. There are some alarming trends occurring in our domestic mining industry. The U.S. share of worldwide hardrock minerals exploration dropped 10 percent in 2000 alone.

Also, since the mid 1990s there has been a sharp decline in mining claims. Further, since 1997 there has been a 66 percent decrease in exploration spending in the United States. These companies are investing abroad.

Why are we losing our domestic industry? First, political risk for mining companies is higher in the U.S. than in most other countries, including those nations many in this room would not travel themselves.

Second, regulatory changes and policies that have been initiated over the past decade are destroying our domestic mining industry. The Millsite Opinion, issued in 1997, and the rewrite of the 3809 regulations in 2001, created massive uncertainty for anyone developing a mine plan in the U.S.

Third, land withdrawals such as the California Desert Protection Act and the Roadless Rule, have jeopardized exploration and development of mineral resources adjacent to withdrawn areas and has locked up millions of acres with promising mineral potential. Uncertainty in the permitting process and frivolous litigation are delaying projects for up to ten years.

Finally, an extended crisis in the surety bond industry, has made it virtually impossible for mining companies to obtain reclamation assurance bonds thus making it exceedingly difficult to obtain permits. All of these factors are destroying our domestic mining industry. Once again, our federal lands policies are running counter to America's economic health and national security.

If we want to remain a strong and vibrant economic and military force, we need to evaluate our current minerals policies and develop one based on common sense. We need policies that promote American industries and American jobs.

I thank you all for coming today, and look forward to the testimony of our witnesses.

Mrs. CUBIN. I now would like to recognize our Ranking Member, Mr. Kind, for any statement that he may have.

STATEMENT OF THE HON. RON KIND, A REPRESENTATIVE IN CONGRESS FROM THE STATE OF WISCONSIN

Mr. KIND. Thank you, Madam Chair. I will try to be brief.

First of all I want to welcome the witnesses today, the Director of USGS, Chip Groat, I appreciate your attendance and the work that you and everyone at USGS has been doing on behalf of our Nation. Thanks for coming, as well as the other witnesses.

And I commend the Chair for holding this very important hearing. I think it is something we need to delve into and keep an eye on from time to time. It is an issue, obviously, of geopolitical significance for our Nation, capabilities abroad and access to supply overseas as well.

But I also believe that the central question to the Subcommittee must be what role can our public lands play in securing our strategic and critical mineral supplies for the future? This is not simply a question of whether the United States contains such minerals. Oftentimes we don't, not in the size, the quantity that we need. It has been typically we don't have them, we have to seek them abroad, and what can we do to ensure that supply line. But inevitably it is also a question of economic viability based on what we do have available and how do we obtain it in a cost effective but also an environmentally minded fashion?

As witnesses will testify today, based on some of the written comments that you have already submitted, most strategic metal ores found in the United States are of such low quality that it is far more economical to buy them abroad. There are some exceptions, two mines, the beryllium operation in Utah and the platinum group of metals mine in Montana stand out. Consequently, our Nation has been and will continue to be dependent on various foreign supplies of strategic metals for the foreseeable future, not unlike what Chairman Greenspan was testifying just a couple of short weeks ago, that it is not all together bad that we try to diversify our sources of natural gas, recognizing we have natural gas supplies here but we are going to be dependent on imports from other countries, and the more diversified we can get, the less dependent we will be on supply disruptions.

There are those that assert the United States should encourage and facilitate the development of our public domain for these critical metals, while others say the geology is simply inadequate for these types of production and would not want the environmental consequences that would come with such development. Most experts would agree, however, that the bulk of the mineral supply of the United States imports for national defense purposes cannot be economically produced in the United States, whether it is from a lack of mineral resources or facilities needed to process the raw ore. Nevertheless, today's hearing will inevitably focus on increased domestic production on Federal lands, but before we consider that option to any great extent, we must also look at the current state of the hard rock mining industry in the United States. The hard rock mining industry is the Nation's most toxic polluter according to EPA's 2003 toxic release inventory. Moreover, the General Mining

Law of 1872 has remained largely unchanged since President Grant signed it into law, and the industry remains exempt from paying royalties on minerals extracted on public lands, nor are completely responsible for the environmental consequences that come from production on these public lands.

My colleague, the Ranking Member of the full Committee here, Mr. Rahall, has once again introduced bipartisan mining reform law of which I am an original cosponsor and which we hope will receive serious consideration by this Committee and by the Congress as a whole.

As with the supply debates and other matters under the jurisdiction of this Subcommittee, a discussion of increased domestic hard rock minerals production on public lands is not complete without considering alternatives to activities that have been potentially hazardous environmental implications. Conservation of minerals through recycling is one area where technological advancements can offset concerns about new production increasing supply.

I am looking forward to perhaps hearing some testimony in regards to what the industry itself is doing in order to reduce the dependence on some of these strategic minerals and sources, and there has been some market forces at play there. Also interested in hearing from any of the witnesses, your insight on recycling efforts and if more can be done in this area in order to reduce our foreign dependence on these crucial supplies.

So once again I commend the Chair for holding the hearing. I thank the witnesses for attendance. I will apologize ahead of time because I have a meeting with Ambassador Zoellick to talk about some trade issues, that I am going to have to step out for, but hopefully return in short order.

Thank you, Madam Chair.

[The prepared statement of Mr. Kind follows:]

Statement of The Honorable Ron Kind, Ranking Democrat, Subcommittee on Energy and Mineral Resources

Ongoing U.S. military activity abroad along with nationwide efforts to ensure our homeland security here at home has drawn the attention of our Subcommittee to the issue of the Nation's strategic and critical minerals supply.

To understand the issue, it is useful to consider the history of the Nation's critical and strategic mineral supply and its relation to geopolitics.

From the end of World War I through the end of the Cold War, the military was concerned about U.S. dependence on foreign nations for supply of certain metallic minerals necessary for military equipment and operations.

Today, the situation is different. The Cold War is over. The nations of the former Soviet Union and Africa are more stable and economically reliant on exports of these strategic minerals.

Technology and geological knowledge have dramatically increased, causing world reserves of these minerals to grow, at times by as much as 700 percent.

In response to these trends, Congress has authorized the sale of many of the Nation's stockpiled minerals, siting the availability of world markets to provide necessary supplies. At this time, only beryllium, mica, and quartz crystals are actively stored for future use.

So, today, the Subcommittee asks the question, "what role can our public lands play in securing our strategic and critical minerals supply for the future?" This is not simply a question of whether the United States contains such minerals. Inevitably, the question of economic viability comes into play.

And, as witnesses will testify today, most strategic metal ores found in the United States are of such low quality that it is far more economical to buy them abroad. There are some exceptions—as evidenced by two mines: a beryllium operation in Utah and a platinum group metals mine in Montana.

Consequently, our Nation has been and will continue to be dependent on various foreign supplies of strategic metals beyond the foreseeable future. As in the case made by Fed Chairman Greenspan for multiple sources of natural gas, from a geopolitical perspective, this is not a bad thing.

There are those that assert the United States should encourage and facilitate the development of our public domain for these critical metals. While others say the geology is simply inadequate for these types of production and would not warrant the environmental consequences.

Most experts would agree, however, that the bulk of the mineral supply that the U.S. imports for national defense purposes cannot be economically produced in the United States, whether it is from a lack of mineral resources or facilities needed to process the raw ore.

Nevertheless, today's hearing will inevitably focus on increased domestic production on federal lands. But before we consider such an option, we must first look at the current state of the hardrock mining industry in the United States.

The hardrock mining industry is the nation's most toxic polluter according to the EPA's 2003 Toxics Release Inventory. Moreover, since the General Mining Law of 1872 has remained largely unchanged since President Ulysses S. Grant signed it into law, the industry remains exempt from paying royalties on the minerals it extracts from the public domain. In comparison, coal, oil and gas producers are all required to pay 8 to 12 percent royalties on production from federal leases. To put this into perspective, since 1872, more than $245 billion in metals and minerals has been extracted without any payment to the owners.

My colleague, the Ranking Member of the Resources Committee from West Virginia, Mr. Rahall, has, once again, introduced a mining law reform bill this Congress, of which I am an original co-sponsor.

Mr. Rahall's bill is designed to bring the Mining Law of 1872 into the 21st Century. If enacted, H.R. 2141, "The Mineral Exploration and Development Act of 2003," would permanently abolish the use of patents in hardrock mining, place an 8 percent royalty on minerals extracted from federal lands, create an abandoned mine lands fund to reclaim abandoned mines, and proscribe new operation and reclamation standards for operators.

As with supply debates in other matters under the jurisdiction of the Subcommittee, a discussion of increased domestic hardrock minerals production on public lands is not complete without considering alternatives to activities that have potentially hazardous environmental implications.

Conservation of minerals through recycling is one area where technological advancements can offset concerns about new production and increasing supply.

I am glad to see Mr. Robert Noel has come to testify on behalf of the Metals Affordability Initiative, a consortium of aircraft and engine manufacturers and key material and component-supplier companies that, among other things, works to reduce the amount of metal used in military and aerospace products.

In addition to industry efforts, Congress must work harder to encourage the recycling of these critical metals.

Whether it is for increased domestic supply, national security, environmental protection, or political purposes, continued research and development of recycling technologies can help us maintain our strategic and critical metals stockpiles and preserve our scenic public lands and waters for future generations to enjoy.

I would like to thank the panelists for their presence today and I look forward to hearing your testimony.

Mrs. CUBIN. Thank you, Mr. Kind.

Would any other members like to make an opening statement? Mr. Gibbons?

Mr. GIBBONS. Madam Chairman, in view of the time and the need to hear from our witnesses, I do have an opening statement, but I will submit it for the record.

[The prepared statement of Mr. Gibbons follows:]

Statement of The Honorable Jim Gibbons, a Representative in Congress from the State of Nevada

Madam Chairwoman, thank you for holding this hearing today to discuss an issue of utmost importance to our Nation's independence, prosperity, economic stability, and most importantly, our safety and security.

We are truly blessed to live in a county rich in natural resources.

As a result of developing our natural resources, not only have we have raised our standard of living exponentially but created the strongest, most technologically advanced military in the world.

Having both a mining and military background, I understand the intricate link between the minerals and metals that our mines extract and the development of the world's most sophisticated technology.

The availability of various materials equates to the most technologically advanced military in the world.

Unfortunately, misinformed environmental groups are incessantly attacking our domestic mining industry.

They work to over-regulate the mining industry to a point where mining operations just cannot bear the burden and flee the United States for business environments that are more fair and predictable.

We have seen this pattern before.

For example, through regulatory burdens we have completely shut off the majority of our public lands from oil drilling, and consequently, we are dependent on foreign nations for this resource.

Because oil is an absolute necessity, we often must ally with countries that do not have our best interests in mind or with nations that are politically unstable.

This puts us in a very vulnerable position.

It is our responsibility as policy makers to ensure that we do not willfully put this country in a position in which we are dependent on other nations for resources that are an absolute necessity for our safety-resources that can be produced domestically.

Furthermore, it is our job to ensure that we protect the basis of our economy which is unarguably our domestic mining industry.

I look forward to hearing from our witness today and I hope that as a result of each of your recommendations, we can make policy which will promote our domestic mining industry and further advancements in technology.

Mrs. CUBIN. Thank you. All opening statements will be submitted.

Now, I would like to recognize the first panel, Dr. Charles Groat, the Director of the United States Geological Survey with the U.S. Department of the Interior. Would you please approach the table?

It is policy for this interior Committee to swear all of the witnesses, so would you mind rising and raising your right hand.

[Witness sworn.]

Mrs. CUBIN. Thank you. The Chairman now recognizes Mr. Groat to testify for 5 minutes. The timing lights on the table will indicate when your time has concluded, and if you are not able to present your entire testimony orally, then your entire testimony will be included in the record.

STATEMENT OF CHARLES G. GROAT, DIRECTOR, U.S. GEOLOGICAL SURVEY, U.S. DEPARTMENT OF THE INTERIOR

Mr. GROAT. Thank you, Madam Chairman, and thank you for the opportunity to testify. I will submit my full testimony for the record and just paraphrase and run through the high points.

I was asked to comment on the U.S. Government's involvement in critical and strategic minerals from a historical point of view to kind of set the framework for how we have been dealing with those.

If we really want to go back to true history, we can go back all the way to Lewis and Clark, when 200 years ago President Jefferson wrote to Merriwether Lewis concerning exploration in the Pacific Northwest and told him to observe, quote, "mineral production of every kind." So the Government has recognized for a long time that mineral resources are a critical part of its economy and of its well-being. Unfortunately, in the early parts of the 19th century the ability to understand those resources, their production and their likelihood for future development wasn't very great because it was spread around in different kinds of reporting mechanisms, so in 1879 the Congress created the U.S. Geological Survey and consolidated much of the activity that it was carrying out in the area of minerals within the USGS. One of the first things the USGS did was to set up a mining statistics division, so that the flow of information about mineral resources in the United States would be adequate.

The creation of the Bureau of Mines in 1925, that function was transferred to them, and then with the abolition of the Bureau of Mines in '96, it was turned back to the United States Geological Survey.

If we look historically at the concern of the Government and its activities in critical and strategic minerals, we really see an ebb and flow of interest and activity. After World War I it was pretty clear that the United States was not going to be and was not at the present time self-sufficient in all the minerals it needed, and by the time World War II came along, the country was really clear in its concern about vulnerability to disruptions in the supply of critical commodities.

As we moved into the atomic age, by that time the United States had become a net importer of many minerals, and it was evidence to all in the industry and those concerned about our economy that this trend posed serious threats for this country in many ways.

In response to that concern the Federal Government took a number of steps to address that. One of them was to create a loan program to encourage exploration on public lands, and second, created the stockpile that has continued to the present of critical and strategic minerals, and put together the Paley Commission, which in '52 produced a report that documented the Nation's mineral position, and developed recommendations about how we ought to deal with our mineral problems.

The ebb comes up again following the Korean Conflict. By that time it was perceived that we were in a surplus situation in most cases, and the interest in doing anything of consequence about critical and strategic minerals wasn't very high.

But there again the flow picks up again as concerns about the Soviet Union and its expanding influence around the globe grew, and there were concerns that we might even end up in a resource war in a sense with the Soviet Union, and even more so, the fact that they might deny us access to some of the foreign supplies, that we were becoming increasingly reliant on, of critical minerals as they extended their influence.

So that at that time, we published in 1973 the first general assessment of the mineral resources of the United States that covered the whole country. This was the first one in 20 years since the Paley Commission, and reported on geologic studies and estimates about that resource base.

In the '70's we tried to develop an approach to helping the country in a probabilistic sort of way understand what the potential was for undiscovered resources. It is one thing to chronicle what is there and what we know about. The question always remains what is the potential of this country to develop more resources? So in

1973 we did such a study, and at that time in the '90's by the time that was completed, we concluded, and I will quote in a sense, that in this first ever report to determine the potential undiscovered deposits of gold, silver, copper, lead and zinc, we concluded, quote, "That there is every reason to believe that, for conventional-type deposits that contain gold, silver, copper, lead, zinc, about as much left to be discovered in the United States as has already been discovered." And we reported what the estimates of those total resources were. This kind of activity will continue to be updated as conditions surrounding the economics and the accessibility to those deposits or potential deposits continues.

In recent concerns, the cold war ebbed in a sense those concerns and we had much less interest in Government action in this sort of thing. Nonetheless, the United States continues to import 100 percent of some very critical resources, indium, manganese, vanadium. We are highly dependent on foreign sources for chromium, cobalt, platinum group minerals and tantalum. In addition, we are increasingly depending on some resources that we at one time were significant producers of, and so we are importing significant amounts of beryllium, copper, lead, lithium, magnesium metal, rare earths and titanium.

Globalization, and I will close with some comments about that. Clearly it is a global market in many ways, and the same is true, as you have pointed out, with our critical and strategic minerals. We are seeing patterns that increase the understanding of our dependence on global supplies. We tried to be a participant by initiating in 2002 the first global assessment of the potential for undiscovered resources which will help us guide policy in our development and ability to use those kinds of resources in a global sense.

Finally, let me close as I started, with the importance of information. It is clear that we need sources of objective, unbiased information about existing resources and the potential for additional resources on both a national and global scale, and as we do that the USGS hopes that it can continue to play that role in providing you, to the private sector, other parts of Government, important information about existing and potential mineral supplies, the information which will help shape a reasonable domestic and global policy on mineral resources.

Thank you, Madam Chairman. Be happy to answer questions.

[The prepared statement of Mr. Groat follows:]

Statement of Dr. Charles G. Groat, Director, U.S. Geological Survey, U.S. Department of the Interior

Madam Chairman and distinguished Members of the Subcommittee, thank you for the opportunity to participate in this hearing and to discuss the role of strategic and critical minerals in our national and economic security. The broad importance of these minerals is often overlooked and misunderstood. Minerals are important to our security and economy. They are the stuff of which our material infrastructure is built.

There is a common misperception that minerals can be found anywhere and that there will never be a problem with sources of these fundamental commodities. I'd like to talk with you today about the U.S. history of mineral supply-and-demand issues and the work the U.S. Geological Survey (USGS) does to provide the Nation's policymakers with reliable, current information that helps sustain the economy and maintain security.

The United States government has a long history of concern about access to the minerals necessary to the functioning of its economy and maintaining a strong national defense. Two hundred years ago, when President Jefferson wrote Merriwether Lewis concerning his exploration mission to the Pacific Northwest with William Clark, he instructed them to observe "mineral production of every kind." Twenty years later, in 1833, George Featherstonhaugh, an English-born geologist, wrote a letter to Secretary of War Lewis Cass expounding on the benefits of dedicating public funds to expand the Nation's knowledge of its mineral resources. He wrote, "It is difficult to form an estimate of the great disadvantages any country lays under, precluded from a correct knowledge of its own mineral resources; and this remark may be especially applied to the United States."

Congress recognized the importance of strategic and critical minerals when it enacted The Domestic Minerals Program Extension Act of 1953, which remains in force. This Act states, "It is recognized that the continued dependence on overseas sources of supply for strategic or critical minerals and metals during periods of threatening world conflict or political instability within those nations controlling the sources of supply of such materials gravely endangers the present and future economy and security of the United States. It is therefore declared to be the policy of the Congress that each department and agency of the Federal Government charged with responsibilities concerning the discovery, development, production, and acquisition of strategic or critical minerals and metals shall undertake to decrease further and to eliminate where possible the dependency of the United States on overseas sources of supply of each such material."

The difficulty in gaining an understanding of the mineral production of the United States during the first half of the 19th century is the legacy of multiple independent government reports on the mineral resources of individual mining districts. At the close of the 19th century, the United States was largely thought to be abundantly endowed with mineral resources. By the first decade of the 20th century, this view was changing and the country began to focus on managing and conserving its natural resources. In 1879, Congress created a single organization, the U.S. Geological Survey, to provide the Nation with knowledge about its mineral resources and the state of their development. Among its first activities, the USGS created a Mining Statistics Division to collect and disseminate information about the Nation's mineral production and to investigate the geology of and the technology employed at several important mining districts.

This vital mining statistics function was transferred from USGS to the U.S. Bureau of Mines (USBM) in 1925 and returned to USGS by the U.S. Congress in 1996. The mineral statistics program was expanded at the USBM as demand increased for minerals data, particularly by defense and emergency preparedness agencies.

Significant concerns about access to strategic and critical minerals arose after World War I when the Nation recognized that it was not self-sufficient in all of the minerals it needed. The implications of being a net importer of minerals and the need to develop elements of a national mineral policy gained the attention of government decision makers and scholars. The onset of World War II brought home to the country its vulnerability to disruptions in the supply of critical mineral commodities. These concerns were heightened by the dawning of the atomic age, as the United States had become a net importer of many minerals. The rise of a communist government in China and the termination of access to Chinese tungsten, as well as the onset of hostilities in Korea, which was also a significant source of that metal, raised concerns about the vulnerability of the United States to disruptions of minerals supplies.

The Federal Government took a number of actions to address the fears about access to strategic and critical minerals. These actions included creating a program of Government loans to encourage exploration for such minerals (the Grubstake Loan Program of the Defense Production Act of 1950, the Defense Minerals Exploration Administration, and successor programs in the USGS), establishing government stockpiles of a wide variety of minerals, and appointing the President's Materials Policy (Paley) Commission. In 1952 the Commission produced a report that documented the Nation's mineral position and made recommendations to address mineral-supply problems.

Following the Korean Conflict, however, minerals were in surplus rather than shortage, and public interest in assuring sources of mineral supplies ebbed. By 1956, even uranium was in oversupply, thanks in part to government efforts to spur exploration and production of that mineral.

However, access to minerals remained a concern of the United States throughout the Cold War. In the late 1970s, guerrilla activity in Shaba Province, Republic of Congo (then Zaire), caused cobalt prices to rise precipitously. Heightened tensions with the Soviet Union in the early 1980s fueled apprehensions about a "resource war" and global competition for resources. Some experts cited mineral resources as one of the dominant factors that led the Soviet Union to invade Afghanistan. This view reflected the broader concern that, if the Soviet Union were in a position to do so, it would deny the United States access to foreign supplies of minerals critical to defense systems or to the U.S. economy.

In 1973, USGS published the first overall assessment of mineral resources of the United States since the 1952 Paley Commission report. Planning by the U.S. Government again focused on access to strategic and critical minerals, support for geologic studies of particular strategic and critical minerals, and the maintenance of adequate stockpiles of materials to meet projected national emergencies.

In addition to undertaking geologic studies of strategic and critical minerals, USGS began in the 1970s to develop consistent, probabilistic techniques for estimating potential for as-yet undiscovered deposits of essential minerals. These techniques were first applied in the United States and used to assist Federal land managers in considering mineral values as a part of land management plans. In the 1990s, USGS conducted the first-ever probabilistic assessment of the entire United States to determine potential for undiscovered deposits of gold, silver, copper, lead, and zinc. The study concluded that "there is every reason to believe that, for conventional-type deposits that contain gold, silver, copper, lead, or zinc, about as much is left to be discovered in the United States as has already been discovered" and reported the estimated total resources of these five metals. USGS will update this assessment periodically as changes occur in minerals utilization, adding economic and environmental analyses when feasible.

Recent Developments and the Present Situation

The end of the Cold War and the breakup of the Soviet Union in 1991 resulted in a lessening in concerns about access to strategic and critical minerals and a decrease in the size and composition of stockpiles. This was based on the assumption that in future emergencies, the United States would have ready access to foreign sources of minerals. Recent events have called that assumption into question. The Federal government continues to maintain stocks of a large number of critical mineral materials such as such as bauxite, chromium, cobalt, columbium, diamond, fluorspar, germanium, graphite, iodine, manganese, mica, palladium, platinum, and tantalum.

The United States imports 100% of such important mineral materials as bauxite, columbium, indium, manganese, and vanadium; and is a net importer of chromium, cobalt, platinum-group metals, and tantalum (see tables 1 and 2). In addition, the United States imports an increasing quantity of mineral materials we once exported. Many of these materials are important components of defense systems or are used in technically sophisticated products, including super alloys in jet aircraft, electronic components, such as capacitors for personal computers and cellular telephones, and semi-conductors. Data collected since 1978 demonstrate that the value of imports of mineral materials has increased faster than the value of exports (figure 1).

Just as the end of the Cold War prompted major changes in defense planning and foreign policy, it also marked a major change in global economies, including the United States. Increased globalization will likely increase global interdependence on mineral supplies, as minerals are mined in one country, processed in another country, and turned into manufactured goods in yet other countries.

Information about both our domestic and global mineral resources remains vital to meeting the economic and national security challenges that the Nation faces. In response to this need, USGS provides information on production and consumption of 100 mineral commodities domestically and in 180 countries. At the same time, USGS conducts research and assessments designed to provide a scientific basis for understanding the Nation's domestic and global mineral resource position. The information provided by USGS is a public good, providing valuable information to market participants that would not be obtained in a private market. Such information is also important should foreign sources become prohibitively expensive in a time of crisis.

In 2002, USGS scientists began a USGS-led, internationally coordinated project to assess potential for undiscovered nonfuel mineral resources on a global scale. The primary objectives of this multi-year project are to outline the principal land areas in the world that have potential for selected undiscovered mineral resources and to estimate the probable amounts of those resources to a depth of 1 kilometer below the Earth's surface. The first priority for the project has been identifying and formalizing relations with other countries and multinational organizations around the world. In addition, USGS is preparing reports on regional geology, recent exploration, significant mines and mineral resources, major past and current production, and supply-demand conditions. These reports will be available beginning in early 2004.

Future Concerns

As developing nations grow, demand on known resources will increase rapidly. For example, among the most dramatic recent changes has been the emergence of the Peoples Republic of China as a major participant in minerals markets. China currently supplies the United States a large number of mineral commodities including: antimony, barite, fluorspar, graphite, indium, magnesium, niobium, rare earths, tantalum, tin, tungsten, and yttrium. However, China's internal consumption of minerals is rising rapidly. China's consumption of copper recently exceeded 1 million tons per year and China will likely be the largest consumer of copper in the world before 2020. China is changing from a country that exports many minerals to one that imports increasing amounts and varieties of minerals. As China and other developing nations grow, trade balances in many mineral materials will shift. The U.S. needs to anticipate these shifts and be prepared with long-term strategies.

Many organizations and agencies need information concerning mineral resources provided by USGS. These organizations include: land management agencies, the Federal Reserve Board, numerous Department of Commerce agencies, and the Departments of State and Defense. Private sector groups, such as industry trade organizations and non-governmental agencies, are also frequent customers and partners. In closing, I would like to reiterate how important minerals are to our security and our economy. They are the stuff of which our material infrastructure is built.

Thank you for this opportunity to testify. I will be pleased to respond to any questions you may have.

Table 1. Percent import reliance and annual US consumpion for selected metallic minerals.

		Percent Import Dependence						U.S. Consumption (tons)
Mineral Commodity	1975	1980	1985	1990	1995	2000	2002	2002e[1],[3]
Aluminum	E[5]	E	16	E	23	33	39	6400000
Antimony	49	48	W	51	75	84	41	18000
Arsenic	W	W	90	100	100	100	100	25000
Bauxite and Alumina	91	94	96	98	99	100	100	3300000
Beryllium	W	W	20	E	E	37	55	220
Bismuth	W	W	W	W	W	95	95	2000[2]
Cadmium	41	55	57	46	E	6	E	560
Cesium	100	NA	100	100	100	100	100	a few tons
Chromium	91	91	75	79	80	67	63	413000
Cobalt	98	93	94	84	82	78	75	10800
Columbium (Niobium)	100	100	100	100	100	100	100	4300
Copper	E	14	28	3	7	37	37	2700000
Gallium	W	56	NA	NA	NA	NA	NA	21000[2]
Germanium	NA	NA	NA	NA	NA	NA	NA	28
Gold	52	18	46	NA	E	E	E	170[2]
Hafnium	W	W	W	NA	NA	NA	NA	NA
Indium	NA	NA	NA	NA	NA	100	100	75
Iron Ore	30	25	21	21	14	10	11	56000000
Iron and Steel	9	15	22	13	21	18	14	107000000
Iron and Steel Scrap	E	E	E	E	E	E	E	72000000[2]
Lead	11	E	13	3	17	18	18	1580000
Lithium	E	E	E	E	E	>50	>50	1600
Magnesium Metal	NA	E	E	E	E	43	54	120000
Manganese	98	98	100	100	100	100	100	700000
Mercury	69	49	51	W	W	NA	NA	NA
Molybdenum	E	E	E	E	E	E	E	19900
Nickel	72	73	72	72	59	56	43	195600[2]
Platinum Group Metals	83	88	92	88				
Palladium						84	69	90 [4]
Platinum						78	93	70 [4]
Rare Earths	E	NA	E	21	6	71	53	10600
Rhenium	67	W	W	W	W	57	59	23
Rubidium	NA	NA	100	100	100	100	100	a few tons
Scandium	100	NA	NA	NA	NA	100	100	NA
Selenium	66	59	W	46	38	W	W	W
Silicon	6	8	25	29	35	47	47	509000
Silver	30	7	60	NA	13	43	61	5340
Tantalum	81	90	89	86	80	80	80	525
Tellurium	50	W	W	W	NA	NA	NA	NA
Thallium	69	W	100	100	100	100	100	0.5
Thorium (ThO2)	NA	NA	NA	NA	100	NA	NA	NA
Tin	84	79	72	71	84	88	79	54000
Titanium Metal	W	W	E	E	36	72	54	19000
Tungsten	55	53	68	73	84	66	70	12900
Vanadium	38	17	W	W	W	100	100	3300[2]
Yttrium	NA	W	100	NA	100	100	100	450
Zinc refined	61	60	70	51	71	72	73	1244000
Zinc all forms				37	35	60	60	1500000
Zirconium	W	W	W	6	W	W	W	W

W - withheld to avoid disclosing proprietary data; NA - not available

[1] apparent consumption unless otherwise noted.

[2] reported.

[3] estimated.

[4] data from Johnson Matthey, Platinum 1998, Interim Review.

[5] Net exports.

Table 2. Sources of mineral imports, 1998-2001.

Mineral Commodity	Percentage of Imports 1 Country	2 Countries	Leading Source of Imports 1st	2nd
Aluminum	60	78	Canada	Russia
Antimony	52	73	China	Mexico
Arsenic	84	93	China	Japan
Bauxite and alumina	29	50	Australia	Guinea
Beryllium	37	53	Kazakhstan	Russia
Bismuth	32	55	Belgium	Mexico
Cadmium	39	68	Canada	Australia
Cesium			Canada	
Chromium	50	70	South Africa	Kazakhstan
Cobalt	23	42	Finland	Norway
Columbium (niobium)	74	83	Brazil	Canada
Copper	30	53	Canada	Chile
Gallium	44	60	France	Kazakhstan
Germanium	35	67	China	Belgium
Gold	46	60	Canada	Brazil
Hafnium	80	88	France	Germany
Indium	40	70	China	Canada
Iron ore	49	87	Canada	Brazil
Iron and steel	19	33	European Union	Canada
Lead	60	71	Canada	Mexico
Lithium	80	96	Chile	Argentina
Magnesium metal	42	62	Canada	China
Manganese	31	52	South Africa	Gabon
Mercury	30	45	United Kingdom	Chile
Molybdenum	30	59	China	Mexico
Nickel	40	53	Canada	Norway
Palladium	44	58	Russia	South Africa
Platinum	55	70	South Africa	United Kingdom
Rare earths	66	93	China	France
Rhenium	61	76	Chile	Kazakhstan
Rubidium			Canada	
Selenium	36	67	Canada	The Philippines
Silicon	23	40	Norway	South Africa
Silver	40	77	Canada	Mexico
Tantalum	49	59	Australia	China
Tellurium	28	56	United Kingdom	The Philippines
Thalium	79	93	Belgium	Canada
Tin	27	50	Peru	China
Titanium metal	36	72	Japan	Russia
Tungsten	48	64	China	Russia
Yttrium	58	94	China	Japan
Zinc	54	66	Canada	Mexico
Zirconium	53	99	Australia	South Africa

U.S. Exports and Imports of Raw and Processed Materials of Mineral Origin

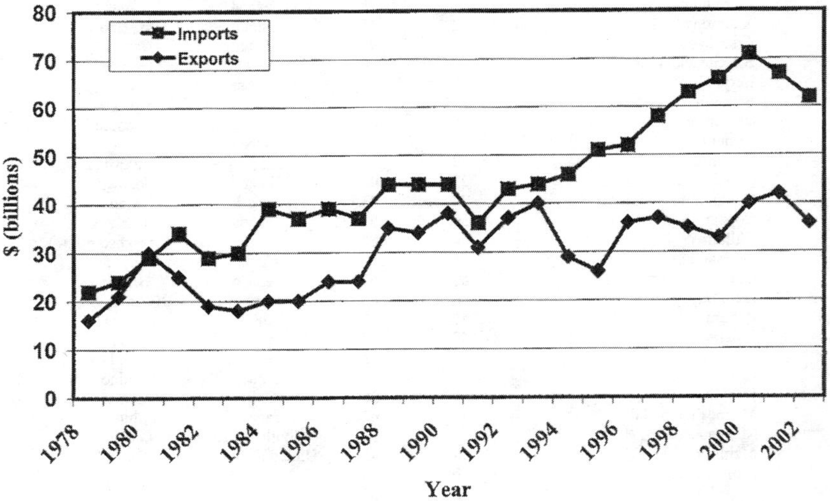

Mrs. CUBIN. Thank you, Dr. Groat. The Minerals Information Team collects and disseminates information on the domestic and international supply and demand of minerals essential to the U.S. economy and national security. Are there other sources of this information, and how is the information used by the Federal Government and the private sector?

Mr. GROAT. I think clearly the most comprehensive source of that kind of information comes from the Mineral Information Team. We report on over a hundred commodities in 80 different countries around the world. While the function is with the Bureau of Mines and with us, it has been recognized as probably the single most public authoritative source. There are estimates made by various consulting firms and foundations about mineral resource potential, but I think that remains the most commonly used, and it has access to the most forms of data, because as a neutral source, companies and others are willing to provide information sometimes on a confidential basis that can be aggregated and reported to an organization like the Bureau of Mines at one time and us now that they might not report to the private firms that they would have some concern about.

Mrs. CUBIN. What is the current budget for the Minerals Information Team?

Mr. GROAT. Let me ask Kate Johnson, who heads our minerals program. Do you have that number, Kate? $16 million, Madam Chairman.

Mrs. CUBIN. And that is the 2003?

Mr. GROAT. 2003 enacted.

Mrs. CUBIN. Right. The fiscal year 2004 budget proposes $750,000 cut in the program. Considering the importance of the program to our national security, why were these cuts proposed?

Mr. GROAT. Well, we had to make some difficult choices in meeting our targets with the budget and we had some significant cuts to be taken in various areas. Mineral Information Team shared in some of those cuts. We were not about to remove that program. We recognized its importance, but we are seeking economies where possible.

Mrs. CUBIN. I just can't help but wonder if the administration, if there is a break in the thought process with some people in the administration in that we need energy for our national security and we need these minerals for our national security. The Interior Department is one of the only agencies that actually brings revenue into the Government, and yet the Interior budget was hit pretty hard. It doesn't make sense to me. Do you have any remarks on that?

Mr. GROAT. I can comment on the USGS aspects of that, and I think what we have learned is that it is important for us to do as good a job as we can in helping people including administrations and constituents and folks that use information, appreciate the importance of the kind of information in the sense of the Mineral Information Program that we develop, and to the degree that we can increase that appreciation, as you and this Subcommittee try to do for mineral resources in general, expand the appreciation for the importance of the resource and information about it, hopefully those kinds of decisions won't have to be made in the future and the country will support them to the degree that you and I seem to think they should be.

Mrs. CUBIN. In what daily operational or concrete ways will that $750,000 affect USGS, or the Information Team I should say?

Mr. GROAT. I don't think at that level of cut that we will stop doing anything that is currently done in the sense of what we report on, the countries that we report on and that sort of thing. I think we will be forced to seek some economies in how we do that. Whether that will mean sacrifices in the quality of data, I certainly hope not, but we will do everything we can to make sure that what we are dependent on for good information continues to be provided.

Mrs. CUBIN. In your opinion, are our current policies toward access to Federal lands affecting our ability to produce strategic and critical minerals?

Mr. GROAT. I think you probably, Madam Chairman, with all due respect, got the wrong guy to be asking the question about access. We are in the science business, the resource estimation business, and don't really get involved in policy questions regarding access. We certainly do get involved in assessing potential resource on Federal lands and recognize that there is certainly potential there, but decisions about whether or not there should be access to them is beyond my scope.

Mrs. CUBIN. I wasn't asking whether or not there should be access. I guess I was asking more about the presence of minerals on the lands, and obviously there is some reason that they are not being explored. You are right, that is not in your job definition, but it is in mine. So your testimony says that there are minerals out there that could be, all things being right, that could be produced.

Mr. GROAT. We certainly can say that, based on our geologic work, that there are potential mineral resources under Federal land surfaces.

Mrs. CUBIN. Thank you.

I would like to recognize Mr. Kind.

Mr. KIND. Thank you, Madam Chair.

And thank you again, Director Groat, for your testimony, your presence here today. Unfortunately, I think the budgetary decisions that you and your department have had to make are consequences of the fiscal policies being pursued. I mean it is the classic tradeoff between the tax cuts that are being enacted right now and the critical resources that are needed in order for us to develop these types of strategic plans for the sake of national security in our country, and you have to deal with the fact that we are operating under very tough, tight, budgetary circumstances, a deficit this year of $460 billion and projected deficits for many years to come. I think, unfortunately, it is going to put a further squeeze on your department and other agencies in your ability to meet the needs that we are asking of you. It is an unfortunate situation, but it is a reality of the fiscal management right now that we see here in Washington.

I believe I heard that you have initiated a global study on strategic minerals that started in 2002; is that correct?

Mr. GROAT. Yes, I did.

Mr. KIND. When do you expect to complete that?

Mr. GROAT. That will be a multiple year effort, as was our global assessment of undiscovered conventional energy resources, and it is intended to be based on the same sort of pattern where we have a group of experts that work collaboratively with comparable experts in other parts of the world and receive information from a variety of sources, and try, in the sequence of groups of critical minerals, to proceed through. We are starting with one group now, and I think, Kate, that total program is estimated to last at least 6 years, isn't it? Yes, 2008.

Mr. KIND. Will there be period reports or reports updating the progress of the study that we can have access to?

Mr. GROAT. Yes, we will release reports as they are generated in sequence rather than waiting to the end, and then do updates as well.

Mr. KIND. I know it is very new. You have just embarked on this global study, but are there any red flags presenting themselves right now that this Committee and this Congress needs to be aware of?

Mr. GROAT. I don't think red flags. I think we will find, as industry is well aware and as we believe we are as well, that the distribution of most mineral resources around the globe is adequate for consumptive use for some time to come. The real challenges get to be in whether they are geologically or in an engineering sense accessible. And as you have struggled with questions, how access to them is available, whether for political, physical or other means. So we hope to be able to identify potential for the resource being there within a kilometer's depth, which in most cases would be accessible, and then that will allow it to be overprinted with other interests and issues about access and availability.

Mr. KIND. Your agency will do an assessment on all that though, not only the presence of these minerals in foreign lands but also the geopolitical consequences, our foreign policy, our relationship with these nations and the potential access that we may have?

Mr. GROAT. No. We will just deal with the resource potential estimates and then with characterization of those resources. Some of the characterizations may affect environmental aspects related to their development, but we won't get into the geopolitical aspect.

Mr. KIND. In your testimony that you submitted there is a Table 1 listing the import reliance on a lot of the minerals, which is quite a variety. The question I have is how many of these minerals are currently available in the United States at significant levels where we could further explore and produce them and reduce our reliance on the imports of these minerals?

Mr. GROAT. I think there are really a couple categories of minerals that you see high reliability on imports. One is those in which this country has very little potential, particularly some of the alloy metals in the steel business, some of the specialty metals, where the geology of this country just isn't really suited for them to be there, and then the other is where economic or other kinds of conditions make it cheaper or more prudent in a sense to obtain them from foreign sources, so it is a mixture of those sorts of things.

We can provide you for the record, Mr. Kind, if you would like, some assessment of which of those is which.

Mr. KIND. Again, just to reiterate your testimony, we basically have a good grip on what mineral sources we have available domestically here; it is not a question of trying to explore or discover new mineral holds in some areas that we are not familiar with; is that right?

Mr. GROAT. Well, I wouldn't say that we have done a uniform assessment job for understanding potential for new resources, particularly in Alaska, for example, where there are vast areas that are inadequately geologically mapped, inadequately assessed for the potential for minerals. There are pieces of territory like that that we can't say that for. There are other parts where we have and others have done a lot of work, and we have a fairer degree of certainty what at least the potential is, but geology always is full of surprises, and if the conditions are right, there is always a good chance of geologists and mining engineers proving that they will find something.

Mr. KIND. Got it. Thank you.

Thank you, Madam Chair.

Mrs. CUBIN. Before I recognize Mr. Gibbons for his questioning, I have to remind the gentleman from Wisconsin that the $20 billion last night that he voted to send to Africa would have gone a long way in meeting the needs of the Interior Department.

Mr. KIND. [Off microphone.]

Mrs. CUBIN. Yes, you did.

I would like to recognize Mr. Gibbons now for his questioning.

Mr. KIND. I am glad someone is paying attention.

Mr. GIBBONS. Thank you, Madam Chairman.

Dr. Groat, thank you very much for your testimony here today. Of the selected metallic minerals that you have listed in your appendix to your remarks, which ones of those are strategic?

Mr. GROAT. That is a definition that has never really been pinned down. I think you could argue on the one hand that every mineral commodity that we use in any quantity is strategic in the sense that our economy can't survive without them. Strategic is usually kind of hinged around whether we have domestic supplies or access, free access, not free, but open access to their supplies in other places. I would tend to say that they are all strategic in that sense.

Mr. GIBBONS. Would you agree that our mining industry is much like, and our minerals industry, is much like our manufacturing industry; as we tend to rely more on the dependence outside of this country, our own institutional basis for being able to discover, explore and develop minerals here is decreased just as it would if it was a manufacturing base where we lose the institutional and educational value of what we have within our own boundaries?

Mr. GROAT. I think there is no question that that is true, Mr. Gibbons. I think one sign of that in the realm we operate in the science area is as we see the closings of schools of mines at places like the University of Nevada and other places, we see even a falling off of commitment of educating people to prepare themselves for careers in that critical area, so that is going to hurt us as well.

Mr. GIBBONS. I noticed when Mr. Kind asked a question you were quick to say that we haven't yet looked at the geology of all of the United States to discover whether these supplies of minerals are adequate or even discoverable. Is it true also to say that we haven't discovered all of the mineral potential yet in the lower 48 that could be of value to us in some of these strategic mineral cases?

Mr. GROAT. No. I think that is definitely true. I think we have a history of proving that, of Carlin gold in Nevada, for example, of new kinds of deposits that are unearthed, and then we look for their potential as something we hadn't looked at before, and there is always that possibility.

Mr. GIBBONS. The geology is an evolutionary cycle and educational evolutionary cycle, and therefore mineral development in these areas would also be subject to new technologies, new educational discoveries about how we produce minerals, how we discover minerals, and how we define minerals as economic for recovery purposes.

Mr. GROAT. I think that is correct, and I think the parallel with our energy resource where technology and increased capabilities have allowed us to extract and process and use beneficially sources or resources that we wouldn't have thought about decades ago, the same thing is true in the mineral industry if we continue to develop the capability to develop that technology and to do the exploration. As you said, that concern that we are not doing that is a legitimate one.

Mr. GIBBONS. It is sort of like water, it seeks its own level, and in other words, where it is easier to find them, we tend to focus our concentration, focus our attention and sponsor where it is easier to find them. If that happens to be overseas, if it is easier to find them, if it is easier to develop them over there, companies are going to go over there.

I was interested, having just returned from Chile, having looked at some of the mines in Chile, because I was constantly hearing the banter from the environmental groups that mining companies are leaving the United States to go to Chile, to go to Third World countries, because they don't have to concern themselves with the environmental issues. May I say that while I was there, I learned a great deal about IOC 14001, which is an international standard that mining companies have to adhere to, which even in those countries they adhere to, which is the same sort of or even greater standards than some of our companies here as well, so that doesn't help.

Real briefly, what minerals do we stockpile?

Mr. GROAT. I don't know. Kate, do we have a list with us?

I don't know, Mr. Gibbons, what the list is.

Mr. GIBBONS. And if we stockpile them, why do we stockpile them?

Mr. GROAT. I can give you a few, a large number of mineral commodities such as bauxite, chromium, cobalt, columbium, diamonds, fluorspar, germanium, graphite, iodine, manganese, mica, palladium, platinum and tantalum. We stockpile them largely because they are commodities for which we are largely dependent on foreign sources, and therefore, should there be some interruption in that supply, that we would have the ability to get by for a reasonable period of time, much as we do with the strategic petroleum reserve in storing oil and gas, oil I mean.

Mr. GIBBONS. Finally, are our stockpiles adequate?

Mr. GROAT. I think that ebbs and flows as does our concern about those minerals. They were depleted after the period of the Korean War and have not been rebuilt up, as far as I know, since that time. Are they adequate to meet a flourishing economy for, I guess the question would be for what period of time? And I think that is where the debate would come, and I don't know the answer to what time they are expected to fill.

Mr. GIBBONS. Thank you, Madam Chairman.

Mrs. CUBIN. Thank you, Mr. Gibbons.

I must correct for the record a statement I made that the amendment to send money to Africa was $10 billion, rather than 20. I want to get that straight for the record.

The Chair would now recognize Mr. Udall if he has any questions for Dr. Groat.

Mr. TOM UDALL. Just a couple of questions. First of all, Dr. Groat, thank you for being here today and thank you for your expertise in this area.

As you are probably aware, New Mexico has a molybdenum mine in Questa, New Mexico, that has been operating for a number of years. I think it is run by Unical. I was just wondering, do we stockpile molybdenum?

Mr. GROAT. I didn't see it on the list. Maybe someone from the industrial panel would know. I don't believe that we do. Molybdenum is one of the steel alloy kinds of minerals that we have been largely self-sufficient in historically, and I would be surprised if we did stockpile it.

Mr. TOM UDALL. So the mine we have in New Mexico supplies you think a major portion of the molybdenum needed in the United States?

Mr. GROAT. I think mines in New Mexico and Colorado have historically played a large role in that, yes.

Mr. TOM UDALL. So we are in a better situation there than some of these other strategic minerals that are there.

Mr. GROAT. Definitely.

Mr. TOM UDALL. I don't have anything else. Yield back. Thank you very much. Thank you.

Mrs. CUBIN. Thank you.

I would like to thank you for your valuable testimony, and tell you that members of this Subcommittee may have some additional questions if they would like. We would submit those to you in writing and hold the record open for 10 days if you could add to those following questions.

Mr. GROAT. Be pleased.

Mrs. CUBIN. Thank you very much.

Mr. GROAT. Thank you, Madam Chairman.

Mrs. CUBIN. And thank you for your testimony, Dr. Groat.

Mrs. CUBIN. I would now like to recognize the second panel to testify.. Mr. Hugh Hanes, Consultant with Brush Wellman, Inc.; Robert J. Noel, Advisor, Metals Affordability Initiative Consortium; Ann Carpenter, Advisor of Women's Mining Coalition; Douglas B. Silver, President of Balfour Holdings, Inc.,

If you would please rise and take the oath.

[Witnesses sworn.]

Mrs. CUBIN. I will once again remind the panel that if they can hold their oral testimony for 5 minutes, their entire testimony will be submitted to the record. I will begin by recognizing Hugh Hanes, Consultant with Brush Wellman, Incorporated.

STATEMENT OF HUGH D. HANES, FASM, CONSULTANT TO BRUSH WELLMAN INC. AND METALS AFFORDABILITY INITIATIVE CONSORTIUM

Mr. HANES. Good morning, Madam Chair and distinguished members of the Subcommittee. I appreciate the opportunity to testify on what I consider to be a critical issue relative to the industrial base of the United States.

The purpose of my testimony will be to demonstrate the linkage between hard rock minerals and strategic and critical metals. I will also give examples where these strategic metals are enabling, both to the aerospace defense and critical civilian infrastructure. And obviously, within 5 minutes I don't have time to deal with all those examples, but they are in the written testimony.

All of these metals, strategic and critical metals are derived from hard rock minerals, whether domestically or of foreign source. These strategic metals are found both in the defense, aerospace and critical civilian infrastructure. The critical civilian infrastructure, at least as I define it, includes automotive, commercial aircraft, computers, telecommunications, electronics, electrical transmission and medical applications. In other words, these are the applications that maintain this country's world leadership as well as the quality of life in this country. These metals have some common characteristics. They are mainly found in high-performance applications. They are used for high reliability because of their unique properties. They become enabling. In other words, they make equipment work better or even work at all. They tend to be pervasive and unrecognized in the infrastructure, and in all cases because they are expensive materials, then tend to be materials of choice.

The manufacturing of beryllium and its alloys is a case with which I am quite familiar, is a case study which demonstrates the independence of mining and specialty metal manufacturing in this country. As was previously noted, beryllium is mined and extracted for minerals in Utah by Brush Wellman, my former company. Many in Congress consider this to be a "western issue," and I use that in quotes. The ore concentrate is shipped to Ohio, our primary production plant. All of a sudden it becomes now an eastern issue. The secondary fabrication occurs in states ranging literally coast-to-coast, Arizona, Massachusetts, New York, Pennsylvania, Rhode Island, and distribution centers around the country and around the world. So it truly is a national issue and it is something that needs to be recognized nationally, not on a regional basis.

Deputy Secretary of Defense, Paul Wolfowitz, when he described to Congress the transformation of the armed services, indicated a list of six goals that the armed services were pursuing, and in that list of six goals the metal beryllium plays an enabling role in equipment and hardware that allows us to accomplish four of those six goals. Then he went further to describe systems under development, and in all cases beryllium played an enabling role. If you look at the Iraqi situation and the weapons that were employed there, beryllium is employed everywhere from strategic surveillance satellites down to the fire control system of the individual tank commander.

Thus, beryllium processing clearly demonstrates that linkage and it also epitomizes what needs to happen within the industrial base.

Precious metals are another example. Between 25 and 55 percent of the so-called noble metals mined and produced domestically, depending on the commodity, are used in critical high reliability electronic applications. For example, they are found in automotive electronics. Computerized ignition systems, automatic transmissions, cruise control devices, anti-lock braking systems, the new generation of electronic suspension systems, are all made better because of the presence of these precious metals, and in effect, society benefits extensively from this because of the increased safety, the increased fleet mileage and decreased emissions in modern automotive equipment.

Silver is another example. Best electrical conductor of all the metals, it finds uses in conductor switches, contacts, circuit breakers and fuses. It not only enhances the quality of life, but it enhances the safety even in our own homes. And as some do, to portray the usage of precious metals as trivial, I think it is totally to ignore the benefits and the increased quality of life that we all enjoy from the judicious application of precious metals in the critical civilian infrastructure.

We support the value of the USGS Minerals Information Team. That has been well covered.

We think that these metals demonstrate the linkage between hard rock minerals and the pervasive use of strategic and critical metals in the domestic industrial base, and we have great concern that a mineral policy not only address the issue of critical minerals, but also the industrial manufacturing base of the critical metals that are derived from these minerals is essential in maintaining the quality of life in this country.

My time is up, Madam Chair, and I am available for questions.

[The prepared statement of Mr. Hanes follows:]

Statement of Hugh D. Hanes, FASM, Consultant to Brush Wellman Inc. and Metals Affordability Initiative Consortium

Good morning Madam Chair and distinguished Members of the Subcommittee. Thank you for the opportunity to testify on this critical issue relative to the industrial base of the United States.

My name is Hugh Hanes. I am a retired Brush Wellman Inc. executive with over 45 years experience in the strategic metals business, including general management of Brush's mining and metallic beryllium operations. Since retirement in December 2000, I have continued as a government affairs consultant to both Brush Wellman and the Metals Affordability Initiative Consortium.

The purpose of my testimony will be to demonstrate the linkage between hardrock minerals and strategic and critical metals. I will also give examples where these strategic metals are enabling to both the aerospace/defense and critical civilian infrastructure.

The importance of these strategic and critical metals is described in the ancient saying, "For Want of a Nail," by some unknown author:

For want of a nail, the shoe was lost,
For want of the shoe, the horse was lost,
For want of the horse, the rider was lost,
For want of the rider, the battle was lost,
For want of the battle, the kingdom was lost,
And all for the want of a nail.

It would be easy to modernize this homily by substituting mineral, metal, turbine engine disk, plane, warfighter, etc. into the text above.

Strategic and Critical Metals: the Hidden Commodities

Strategic and critical metals are often referred to as hidden commodities. Metals availability is usually assumed by those in the user base who are dependent on the specialty aerospace metals industry. Usually, they're only noticed when either they become unavailable, e.g., the cobalt shortage of the 1970's, or a component fails.

However, they have similar characteristics. All of these metals are derived from hardrock minerals. Furthermore, many of these minerals are no longer domestically mined, as is shown in Table 1. As has already been discussed in this hearing, the domestic mining and minerals industry is declining. As my colleague, Mr. Noel, will describe in his testimony that the domestic specialty metals manufacturing base is declining as well.

These strategic metals are found in both the defense/aerospace and critical civilian infrastructure. For the purpose of this testimony, the critical civilian infrastructure can be defined as automotive, commercial aircraft, computers, telecommunications, electronics, electrical transmission, and medical applications. In other words, these are market sectors which help maintain this country's world leadership and quality of life.

In many instances, the applications are hidden, or buried deeply in the systems where these metals perform critical functions. Examples would include beryllium in aerospace/defense systems, precious metals in high-performance electronics, and rare earth metals in electro-optics.

Characteristics of Strategic and Critical Metals

Strategic and critical metals have common characteristics. They are most often found in high-performance applications, where there are requirements for combinations of high temperature resistance, high strength requirements, and corrosion resistance, etc. These classes of metals are used for high reliability, e.g., nickel-based superalloys for aircraft turbine engines, gold-plated connectors in automotive ignition systems, and silver-plated contactors in electrical transmission. Furthermore, they are enabling in their applications, e.g., beryllium optics in surveillance satellites, precious metals in electronic components for computer, and copper beryllium and precious metals in automotive electronics. Usually, they are the most expensive solution, because competitive materials have already been eliminated for non-performance in the particular application.

As shown in Table 2, these metals are pervasive in systems that serve both aerospace/defense and the critical civilian infrastructure. In all cases, they are the material of choice, i.e., they are used because of performance requirements. In most of the cases, they are enabling to the operation of the particular system.

Beryllium and Its Alloys: Case Study

The manufacturing of beryllium and its alloys is a case study which demonstrates the interdependence of mining and specialty metals production. Beryllium is mined and extracted from minerals in Utah by Brush Wellman (a "Western issue"). The ore concentrate is shipped to their primary metals production plant in Ohio (now it becomes an "Eastern issue"). Brush does secondary fabrication of its beryllium products in plants in Arizona, Massachusetts, New York, Pennsylvania, and Rhode Island. They have distribution centers in California, Illinois, Michigan, and New Jersey (it's really a "Domestic issue"). Brush also has distribution centers globally, serving over 5,000 customers for beryllium products globally.

In his testimony to Congress [1], Deputy Secretary Paul Wolfowitz stated, "The Department of Defense is undergoing a substantial transformation of the Armed Services. ... by pursuing a host of transformations including precision, surveillance, networked communications, robotics and information processing." That beryllium is critical to 4 out of 6 of the Secretary's goals can be demonstrated by examples of both current and developmental systems that use beryllium because of its unique properties. Specific examples are shown in Figures 1 to 4.

- Homeland Security—"U.S. forces must protect critical bases of operations and defeat weapons of mass destruction and their means of delivery." Beryllium is a key structural element in both the PAC-3 system and those interceptor systems under development.
- Deny Enemies Sanctuary—"Space denial capabilities, such as ground-based lasers ... require the development and acquisition of robust capabilities to conduct persistent surveillance of vast geographic areas and long-range precision strike." Beryllium is used in long-range surveillance systems, guidance, and is in development as seekers in new missile and ground-based lasers systems.
- Projecting and Sustaining Forces—"increasing U.S. advantages in stealth, standoff, hypersonic and unmanned systems for power projection; and developing ground forces that are lighter, more lethal, more versatile, more survivable, more sustainable, and rapidly deployable." Beryllium is used extensively in reconnaissance satellites, FLIR's, improving stand-off ranges for virtually every new generation targeting device, and battlefield surveillance, including the tank commander's sight on the M1A2 Abrams.
- Enhancing Space Capabilities—"become more dependent on space systems for communications, situational awareness, positioning, navigation, and timing." Applications of beryllium include instruments and critical structures in reconnaissance and surveillance satellites, defense weather satellites such as NPOESS, and the new generation of military communications satellites.

Mr. Wolfowitz goes on to describe systems under development, and in all cases, beryllium plays an enabling role:

- Joint direct attack munitions (JDAM's) and other precision guided munitions
- Stealthy F-22's
- Development of missiles defenses, including the Airborne Laser program
- Enhanced electro-optical capability for Global Hawk and other UCAV upgrades
- Precision weapons—weapons that are precise in time, space, and in their effects
- Missile defense—pursuing parallel technologies to meet the same objectives—for example, the kinetic kill boost vehicle and a space-based laser (beryllium is critical to both concepts)

Thus, beryllium processing clearly demonstrates the linkage between mineral resources in the Western U.S. and metals manufacturing in the Domestic industrial base.

Precious Metals Perform Critical Functions in the Civilian Infrastructure

Precious metals are often portrayed by opponents of hardrock mining as unnecessary metals, but they perform critically enabling functions in the civilian infrastructure. Between 25 and 55% of the so-called noble metals mined and produced domes-

---

[1] "Prepared Statement for the Senate Armed Services Committee Hearing on Military Transformation", by Deputy Secretary of Defense Paul Wolfowitz, April 9, 2002.

tically are used in critical, high-reliability electronic applications because of their combination of oxidation resistance, electrical and thermal conductivity, and their resistance to corrosive environments. These high-reliability requirements dictate the selection of precious metals for many applications in a wide variety of industries, including the electrical, electronics, automotive, telecommunications, semiconductor, computer and medical industries. Examples of typical applications can be found in Table 3 and are illustrated in Figure 5 of this testimony.

Because of their high intrinsic cost, precious metals are often plated or laminated onto base metals to give added strength and to lower the cost of the component. Although gold remains the industry standard in many of these applications, gold and gold alloys as a cover over palladium and palladium-silver alloys are often used.

One of the major uses of high-reliability components containing precious metals can be found in automotive electronics. Under-hood interconnects for computerized ignition systems, mass air flow sensors, automatic transmissions, cruise control devices, anti-lock braking systems, and new generation suspension control systems all are made more reliable by employing precious metal containing components. Society benefits extensively from the use of these electronic components because of the increased safety, increased fleet mileage, and decreased emissions of the modern automobile.

Silver finds many used in both medicine and in electrical transmission. While silver's importance as a bactericide has been documented only since the late 1800's, its use in purification has been known throughout the ages. Silver also has a variety of uses in pharmaceuticals forming the most powerful compounds for burn treatment, for example. Silver is the best electrical conductor of all metals and is hence used in many electrical applications, particularly in conductors, switches, contacts, circuit breakers, and fuses. Thus, silver enhances the quality of life and safety even in our own homes.

To portray the usage of precious metals as trivial, as has been done by opponents of mining, is to totally ignore the benefits and increased quality of life we all enjoy from the judicious application of precious metals in the critical civilian infrastructure.

The Value of the USGS Mineral Information Team

The USGS Mineral Resource Program's Mineral Information Team is the only comprehensive source of statistical data on Mining and mineral commodities both domestically and internationally and is critical to the mining industry and to the nation as a whole. As a net importer of minerals, including many strategic minerals, the United States' ability to develop and implement global mineral-related strategy could be severely compromised without the availability of reports produced by this program. In addition, the analytical expertise of the program's mineral commodity and country specialists is vital to answering mineral related questions of a domestic and an international nature. A loss or reduction in expertise for tracking the world "hot spots" with respect to strategic and critical materials could negatively impact U.S. intelligence and national security. As a world leader, the U.S. must have a comprehensive and essential understanding of the worldwide commodity markets necessary for strategic and critical materials necessary to a healthy economy.

Summary and Conclusions

The purpose of this paper has been to demonstrate the linkage between hardrock minerals and the pervasive use of strategic and critical metals in the Domestic industrial base

1. Strategic and critical metals are derived from hardrock minerals, both domestic and foreign.
2. Component manufacturing is located across the country but primarily in Eastern (non-mineral) states and is dependent on hardrock minerals as the source of primary metals.
3. Both domestic aerospace and defense and critical civilian industries are dependent on a shrinking industrial base for their strategic and critical metals.
4. Continuation of the USGS Mineral Information Team will assure a comprehensive and essential understanding of the worldwide commodity markets necessary for strategic and critical materials necessary to a healthy domestic economy.

A well-conceived minerals and metals policy should protect and encourage maintaining both the development of domestic mineral resources and the strategic and critical metals industry. We have lost or are losing these capabilities as we speak. They have been precipitated by a series of unwise political decisions largely over the last 10 years which discounted the importance of a U.S. minerals base.

I look forward to working with Resource Committee members and my mining colleagues to reconstruct these vital elements of our national infrastructure.

Madam Chair and distinguished Members of the Subcommittee, I sincerely appreciate the opportunity to testify before you and would be glad to answer any questions you may have.

Table 1. U.S. Reliance on Mineral and Metal Imports

100% Reliance Bauxite and Alumina, Columbium, Manganese, Strontium, Yttrium	60-69% Reliance Titanium (Sponge), Tungsten
80-89% Reliance Platinum, Tantalum, Tin	50-59% Reliance Nickel, Silver
70-79% Reliance Chromium, Cobalt, Rare Earths, Titanium Concentrates	< 50% Reliance Aluminum, Beryllium, Copper, Iron Ore, Magnesium

Source: Mineral Commodity Summaries 2001, USGS

Table 2. Strategic and Critical Metals in Defense Systems

System	Percent Metal Content	Primary Metals	Secondary (Alloying) Metals
Airframe and Structures	67% of typical airframe	Titanium, Aluminum	Beryllium, Chromium, Iron, Magnesium, Manganese, Scandium, Silicon, Tin, Vanadium, Zirconium
Turbine Engines	80% of typical engine	Titanium, Aluminum, Nickel	Beryllium, Chromium, Cobalt, Iron, Magnesium, Manganese, Scandium, Silicon, Tin, Vanadium
Space and Missiles	Enabling for space propulsion; critical for structures	Aluminum, Titanium, Nickel, Beryllium	Cobalt, Chromium, Iron, Scandium, Magnesium, Manganese, Silicon, Tin, Vanadium

Source: Air Force Research Laboratory presentation to the House Mining Caucus, July 16, 2002

Table 3. Some Typical Applications for Precious (Noble) Metals in the Critical Civilian Infrastructure

Industrial Sector	Typical Applications
Automotive:	Connectors Terminals Switches Bond Pads Lead Frames for: Air Bags Anti-Lock Brakes Mass Airflow Controls Speed Controls Powered Accessories Sensors
Telecommunications:	Connectors Switching Systems Cellular Phones Shielding Materials
Computers:	Connectors Switches Lead Frames Heat Sinks Multichip Modules Storage media
Medical:	Pharmaceuticals, home health-care equipment
Aircraft:	High-reliability switches, connectors and contactors
Power Transmission:	High-voltage switching gear, conductors, switches, contacts and fuses

Sources: Brush Engineered Materials and the Silver Institute

Mrs. CUBIN. Thank you.

I would now like to recognize Robert J. Noel, Advisor of the Metals Affordability Initiative Consortium.

STATEMENT OF ROBERT J. NOEL, EXECUTIVE ADVISOR TO THE METALS AFFORDABILITY INITIATIVE CONSORTIUM

Mr. NOEL. Good morning, Madam Chair and distinguished members of the Subcommittee. The purpose of my testimony is to discuss aerospace metals, the Metals Affordability Initiative Consortium, and the need for Government support of the industrial base.

The key defense metals are aluminum beryllium, nickel base superalloys and titanium. The unique basic metal properties are further enhanced by key elemental alloying additions. All these metals are derived from hard rock minerals, either domestic or foreign. Metals by weight are 67 percent of our military aircraft structures, and on the average 80 percent of the gas turbine propulsion systems.

The U.S. industrial base that produces these systems consists of supply chain of metal producers, component producers and original equipment manufacturers of both propulsion and aircraft systems. The MAI consortium of 15 companies has manufacturing facilities throughout the United States and is not concentrated in any one region. The aircraft OEMs are Boeing, Northrop Grumman and Lockheed Martin. The engine OEMs are Pratt-Whitney, General Aircraft Engines, Rolls Royce and Honeywell.

In the recent economic conditions the impact has been most severe on the metal and the metal parts producers. In the category of metal producers we have Allegheny Technologies in their Oremet Division, Brush Wellman, Carpenter, Special Metals, Timet; and the component producers are Howmet, PCC and Ladish.

Two examples of consolidation in the industry. One is in the area of titanium which is a very important aerospace metal. The manufacturing of titanium sponge is the initial step in making this crucial metal. In 1990 there were three or less sponge producers with a total capacity of 30,000 metric tons. Today there is one domestic sponge producer, Timet, with a capacity of 8,600 metric tons.

A second example is Ladish, my former employer. It supplies aerospace metal components for gas-driven engines, aircraft and space. Peak employment was 5,300 employees in 1979. Current employment level is less than 800.

In addition to the consolidation that is occurring in these industries, there are other factors. Obviously, there are huge global issues that have resulted in intense competition. There is also foreign government industry investment.

We also have technology issues. Company-funded research and development is declining. We see workforce and skills eroding. We definitely have a concern for the availability of engineering talent to support our industrial base.

There is a perception that the aerospace specialty metals businesses are performing well, but in truth the five metal producers lost money in 2002, and additional losses are expected in 2003. These metal producers supply 90 percent of the specialty metals for the aerospace and defense industry. The other aspect of the financial situation is there is on the average a 35% reduction in R&D expenditures. We feel one of the answers to the problem is emphasis on technology, and that feeling was supported by the Commission on the Future of the U.S. Aerospace Industry.

MAI was a consortium formed in 1998. Our theme is "Using technology innovation to transform and sustain the specialty metals industrial base." The collaborating Government agency is the Air Force Research Laboratories at Wright Patterson Air Force base.

We consider this as a template for Government-industry collaboration. The objective of MAI is to take core technologies through the manufacturing development, and then work with the systems offices to insert those in military systems. The technology work that has been done to date has yielded a return of $650 million on the Government investment. This effort is cost shared by the industry at a minimum rate of 25 percent, and they have contributed $16 million.

MAI currently has 14 active projects with total value Government funded component of 39 million which came largely from congressional additions to the DOD budget. We are currently working with the C-17 systems, the F-15 and C-130 to insert key technologies. We are also working with the key propulsion systems offices to insert technologies as well.

In summary, aluminum beryllium, nickel base superalloys and titanium are essential for U.S. military aircraft and space systems. These metals and their alloying elements are all derived from hard rock minerals.

Both domestic aerospace and defense and critical civilian industries are dependent upon a shrinking industrial base for strategic and critical metal components.

The specialty metals producers showed financial losses in 2002 that are projected to continue in 2003.

The Metals Affordability Initiative Consortium has demonstrated technology successes, a significant return on Government investment, which has come largely from congressional additions to the DOD budget, and is working to insert these technologies in defense systems.

In our opinion, the keys to transforming the specialty metals industrial base are technology innovation, Government-industry collaboration, and a metals policy that includes financial support for core technology programs and technology investment.

Thank you, Madam Chair, and distinguished members of the Subcommittee.

[The prepared statement of Mr. Noel follows:]

Statement of Robert J. Noel, Executive Advisor to the Metals Affordability Initiative Consortium

Good morning Madam Chair and distinguished Members of the Subcommittee. Thank you for the opportunity to testify on this critical issue to the United States Aerospace Specialty Metals Industrial Base.

My name is Bob Noel. I am a retired Ladish Co., Inc. executive with 39 years experience in the specialty metals business with the most recent position being Vice President of Business Development/Technology. I have also served as a Trustee of the Forging Industry Education and Research Foundation since 1989 and am currently Chairman of the Technology Roadmap Committee. Since February of 2002, I have served as Executive Advisor of the Metals Affordability Initiative (MAI) Consortium.

The purpose of my testimony will be to discuss Aerospace metals, U.S. specialty metals industrial base, the importance of technology, and the Metals Affordability Initiative (MAI) Consortium performance, and the need for government support of the industrial base.

Metals and Applications

The key defense metals are aluminum beryllium, nickel base superalloys and titanium. The unique basic metal properties are further enhanced by key elemental alloying additions. All of these metals are derived from hardrock minerals, either domestic or foreign. As Hugh Hanes described in his testimony, metals represent the major portion of U.S. military propulsion, aircraft, and space systems. Typical applications are shown for the F-22 aircraft, F-135 engine and a space rocket engine in Figures 1 though 3.

U.S. Aerospace Specialty Metals Industrial Base

The industrial base that produces military systems consists of a supply chain of metal producers, component producer's, and original equipment manufacturers of propulsion and aircraft systems. The Metals Affordability Initiative (MAI)

Consortium of 15 companies is most representative of the specialty metals industrial base. The member companies are Allegheny Technologies, Boeing, Bush Wellman, Carpenter Technologies, General Electric Aircraft Engines, Honeywell, Howmet Castings, Ladish, Lockheed Martin, Northrop Grumman, Pratt and Whitney, PCC, Rolls Royce, Special Metals and Timet. A minimum of two companies represents every element of the supply chain. These companies have manufacturing facilities through out the United States and are not concentrated in one region. The companies and their position in the supply chain are shown in Figure 4.

Industrial Base Consolidation

The entire specialty metals industrial base is affected by consolidation, downsizing to fit available business volume, globalization, and high capital cost needs. The impact has been most severe on the following metal and parts producers:

Metal Producers

- Allegheny Technologies (Oremet Titanium)
- Brush Wellman
- Carpenter
- Special Metals
- Timet

Component Producers

- Howmet Castings (An Alcoa Business)
- PCC Structurals (Acquired Wyman Gordon)
- Ladish

Titanium is a very important Aerospace metal. Manufacturing of Titanium sponge is the initial step in the metal production process. In 1990 there were three U.S. sponge producers—Timet, RTI and Allegheny Technologies (Oremet) with combined capacity of 30,000 metric tons. Currently there is one domestic sponge producer B Timet with a capacity of 8600 metric tons. The U.S. sponge capacity growth and decline are illustrated in Figure 5.

Ladish is a metal forging producer that was founded in 1905. It forges Aerospace metals for gas turbine engines (Propulsion), aircraft and space applications. Peak employment was 5300 employees in 1979. The current employment level in less than 800. Their business is focused on the high technology segment and is very capital intensive. The 10,000 ton isothermal press shown in Figure 6 is used primarily to produce nickel base superalloys for gas turbine engines and is used to illustrate the high equipment capital needs.

Industrial Base Perspective / Economic Performance

The Aerospace metal industry from 1990 to 2003 was affected by the following conditions:

1. U.S. Aerospace Industrial Base Reductions
   - OEM's reduced with very significant consolidation
   - Metal suppliers and component producers also reduced
2. Global Factors
   - Intense competition
   - Foreign government industry investment
3. Technology Issues
   - Company funded Research and Development declining
   - Workforce/skills eroding with engineering talent availability concern

A general economic perception is that the Aerospace specialty metals businesses are performing well. A review of the metal and component producer's financial performance is illustrated in Table 1. The results show significantly lower stock prices caused by deteriorating financial performance. The five metal producers lost money. These suppliers produce 90% of the specialty metals for aerospace and defense industry. The declining financial performance has resulted in an average 35% reduction in Research and Development expenditures.

In an analysis of the industry and the competitive threats, an essential element is technology innovation. The Commission on the Future on the United States Aerospace Industry findings on the role of technology was "A recurring message we hear from the inputs the commission received is that investments in technology will provide the KEY enablers to our nation's future aerospace capability..."

Metals Affordability Initiative Consortium

The Metals Affordability Initiative (MAI) Consortium was formally started in 1998. The objectives were to provide a source of funding to advance Metals Technology and maintain the U.S. Defense Aerospace Specialty Metals Industrial Base. The theme is "Using Technology innovation to transform and sustain the Specialty

Metals Industrial Base." The collaborating government agency is the Air Force Research Laboratories at Wright Patterson Air Force Base. The MAI consortium has grown to fifteen companies.

The Technology programs have been very successful with a projected return on government investment of over $650-million which exceeds 15 to 1. The government investment is supplemented by a minimum cost share of 25% or $16-million through FY03. A key to the technical success is collaboration. We also consider MAI as the template for Government-Industry Collaboration.

MAI currently has 14 active projects with the government funded component at $39-million, largely from Congressional additions to the DOD budget. The key technology projects with system interest and opportunities are:

- Electron Beam Melting of Titanium Slabs
- Friction Stir Welding
- Laser Additive Manufacturing
- Thin Wall structural Castings
- High Yield Casting of Turbine Airfoils
- Roll Forming of Engine Casings

The objective of MAI core technologies is to take metals and process technology concepts through manufacturing process demonstrations. The mature core programs are demonstrating technical success and meeting business case goals. The next step is insertion into military systems. The four technologies selected for insertion into the C-17 are illustrated in Figure 7. Each technology offers a total systems cost benefit. The application of these technologies is pervasive and can be applied to other systems such as the F-15 and C-130.

Airframe and propulsion systems have different engineering design considerations. Gas turbine engine components operate at higher temperatures and there is significant use of nickel base superalloys. The two processes identified for insertion into the F-135 are shown in Figure 8. The roll forming process can be used for titanium and superalloy cases. The casting process innovation being applied to airfoils will also result in a system performance improvement.

Summary and Conclusions

1. Aluminum, beryllium, nickel base superalloys, and titanium are essential for U.S. military aircraft and space systems. These metals and their alloying elements are all derived from hardrock minerals.
2. The U.S. Specialty Metals industrial base consists of metal producers, component producers, and OEM's with operations located throughout the US.
3. Both domestic aerospace and defense and critical civilian industries are dependent on a shrinking industrial base for their strategic and critical metal components.
4. The specialty metal producers showed financial losses in 2002 that are projected to continue in 2003.
5. The Metal Affordability Initiative (MAI) Consortium has demonstrated technology successes, a significant return on government investment obtained largely from Congressional additions to the DOD budget, and is working to insert the core technologies in defense systems.
6. The keys to transforming/sustaining the U.S. specialty metals industrial base are technology innovation, government-industry collaboration, and a metals policy that includes financial support for core technology programs and technology investment.

Madam Chair and distinguished Members of the Subcommittee, I sincerely appreciate the opportunity to testify before you and would be glad to answer any questions you may have.

Mrs. CUBIN. Thank you, Mr. Noel.

I would like to now recognize Mr. Gibbons to introduce our third witness.

Mr. GIBBONS. Thank you, Madam Chairman.

Indeed, we are all pleased when one of our constituents has an opportunity to visit the Hill, but we are even more pleased when they take time to testify before the Committee, and Ann Carpenter, a constituent of mine from Reno, Nevada has a longstanding association with the mining industry, as advisor to the Women's Mining Coalition, brings I think not only a new perspective but a new view or vantage of how important the mining industry is, not just to the industry or the industrial base but to the population base that we have that works these industries, and I am very pleased to welcome Ann Carpenter to our Committee, and certainly look forward to her testimony.

Thank you, Madam Chairman, for allowing me to say a few words about our next witness.

Ms. CARPENTER. If I can just take a minute before my time starts and we could put the graphs up. I have a couple of graphs that illustrate some alarming trends that I would like to talk about.

Mrs. CUBIN. Could you either raise that or pull it out more so the panel can see it? And if Mr. Renzi and the folks back there would like to move over someplace, they can do that, or you can stay and we won't look at your faces any more.

[Laughter.]

STATEMENT OF ANN S. CARPENTER, CONSULTANT, DOMESTIC AND INTERNATIONAL MINERAL DEVELOPMENT, ADVISOR AND PAST PRESIDENT OF WOMEN'S MINING COALITION

Ms. CARPENTER. Thank you, Madam Chairman, and members of the Committee, for this opportunity to testify.

My name is Ann Carpenter. I am a current Advisor of the Women's Mining Coalition and a past President. This is a grass roots organization with members nationwide, representing the broad spectrum of jobs and the industry diversity.

I am a professional exploration geologist with over 20 years of experience doing mineral and exploration development both here in the U.S. as well as internationally in a number of different companies overseas. My testimony will discuss the alarming downward trends that we see in mineral development here in the U.S., and the long-term impacts.

The downward trends include a drastic decrease in exploration spending since 1997, translating to fewer new discoveries here in the U.S. Both of these figures are compiled from using some USGS data. Figure 1 is on the bottom and it is also page 5 in my testimony. Figure 2 is on the top and that is page 6 in my testimony.

Exploration for mineral resources has continued globally even through the low gold market, but what we do see instead here in the U.S. is a radical decrease in exploration spending here. Investing here in the U.S. is one of steady decline. Our market share, the expenditure of exploration budgets here. In 1997 and pre-1997 they were 30 to 50 percent and they have decreased to less than 10 percent expenditures. In other words, if I spend $10 million in exploration expenditures, I am spending less than a million here and over 9 million overseas.

Why is this occurring? I try to detail in my testimony five different issues that are important to me, but these are not the only issues that are contributing to it. There is a lot of uncertainty in operating here in the U.S., and a lot of it is due to many different factors, but these five I am going to summarize:

Revision to the 3809 regulations started in 1996, taking all the way to 2001; the Millsite opinion further fueled this uncertainty; inconsistencies in the interpretation and implementation of regulations; the bonding crisis; and access issues.

From my personal perspective I was negotiating in 1997 to have a $2 million budget to spend in the U.S. developing mineral projects. The company that I was working with, they were a little bit tenuous because of the revisions of the 3808 regulations, but they were ready to go ahead. As soon as the Millsite opinion came out by the former Solicitor Leshy, they left this country faster than anybody I have ever seen. The negotiations ended, and instead the investing company perceived that it would be less risky from a regulatory perspective to work in Africa than here. Well, I have worked in Africa. I understand risks there, and that spoke volumes to me.

Other uncertainties are wrapped up in access. Conservation restrictions from the Federal and State have grown roughly from about 10 percent in the 1960's to better than 45 percent today, and as much as 55 percent when we include the Roadless Initiative. These impeded mineral development on Federal ground, and a lot of what we see is that these impediments come and these restrictions come without doing full detailed mineral assessments.

I worked overseas, like I have indicated, and there has not been a place that I haven't worked where this one recurring question comes up. "Why are you here taking my minerals?" They understand that there is economic development that goes with it, but it is an important question to me because then I question both our mineral policy and our foreign policy. A lot of times they will answer that question by saying, "We understand you can't really access or permit your resources." But another thing that they make statements around is that they perceive that we are developing minerals there while we are saving ours here at home.

Mining is difficult and mineral resource development is difficult no matter where you go in the world, but at least in other countries, as one senior mining executive says, and a lot of us feel in the field, I know that if I meet the regulatory and the legal requirement, I will get a permit. I do not know that in the U.S. I never know if or when I will finally get a permit even if I can demonstrate that the mine will be in full compliance.

A mineral economist concluded recently that the most recent threat and the most serious threat to the mining industry's long-term sustainability in the U.S. is the regulatory changes in the final 4 years of the Clinton administration, including the revisions of the 3809 regulations and the Millsite opinion.

I implore you, let us work together to reverse these radical decreasing trends, these downward trends. Let us develop and implement a working mineral policy here in the U.S., and let us provide for the national and economic security now and into the future.

I appreciate the time and the opportunity to speak before you. Thank you, Madam Chairman.

[The prepared statement of Ms. Carpenter follows:]

Statement of Ann S. Carpenter, Consultant, Domestic and International Mineral Development, Advisor and Past President, Women's Mining Coalition

Chairman Cubin and members of the Committee:

Thank you for the opportunity to testify before this Subcommittee today. My name is Ann Carpenter. I am an advisor to and past president of the Women's Mining Coalition and work as a professional exploration geologist in the domestic and international mineral development arena. Today I will discuss impediments to mineral exploration and development in the U.S. and impacts to our domestic mining industry. I am here representing the Women's Mining Coalition.

The Women's Mining Coalition was organized in 1993 and is a grassroots coalition supporting environmentally responsible mining. WMC membership is diverse and nationwide, representing many sectors of the mining industry, including coal, energy, metals, construction materials, stone, industrial minerals, and the vendors and manufacturers who provide goods and services to the mining industry.

I have a bachelor's degree in Geology from Montana State University in Bozeman, Montana and have more than 20 years of experience in the mining industry. I have worked throughout the Western U.S.—from Alaska to Southern California and east to Colorado. Additionally I have worked in the international exploration arena—in Mexico, Peru, Argentina, Chile, and Africa—evaluating mineral properties and company acquisition opportunities, and assessing the mineral potential of several countries for various companies. My work experiences have focused on the exploration for and development of metal deposits (gold, silver, copper, lead, zinc), with some additional ventures into the industrial mineral sector.

BACKGROUND

Today I would like to discuss the impediments to mineral exploration and development here in the U.S., and concerns for the long-term viability of our domestic mining industry from the perspective of an exploration geologist. Exploration is the "research and development" arm of the mining industry. It is crucial and necessary in order to keep a flow of projects in the production pipeline, supplying us with the raw materials we need.

The mining industry is an important part of the U.S. industrial base. This industry provides many of the raw materials required for housing, transportation, power generation and transmission, communications, the tech-industry, health care, agriculture, and the arts. Mined materials are also used to create and maintain a clean healthy environment. Mining contributes to the nation's overall standard of living, contributing to the health and well being of not only all Americans, but people the world over.

The U.S. has seen a drastic decrease in the exploration for and development of mineral resources since 1997, not all of which is attributable to decreases in metals prices. Exploration for mineral resources has continued globally even through low metals prices, yet the U.S. has lost market share in mineral development. At an ever increasing rate, investment dollars are being spent on projects overseas instead of here in the U.S. This has occurred even though the U.S. is highly regarded for its diverse geologic terrains and related mineral resource potential.

Exploration for mineral resources is a very risky business—the statistics for success are staggeringly low. Approximately 1 out of 1000 projects reviewed will progress to an advanced-stage exploration and development phase. Making a mine out of an advanced-stage project is dependent on many variables, and mine decisions do not come easily or cheaply.

Investment capital to advance exploration properties generally comes from what is termed the "junior market". Historically, these have been small companies who secure their funding from the venture capital markets in Toronto, Vancouver, London, and other places around the globe. Investors assess properties and investment opportunities based on many factors, two of which are key—the mineral potential of any given area and the political stability of the country where the property is located. Both are considered when reviewing for investment attractiveness (to determine if investment dollars will be well spent). The exploration projects of the small companies today often become the development projects of larger operators tomorrow.

The Fraser Institute (an independent Canadian economic and social research and educational organization) has conducted an annual survey of metal mining companies since 1997, assessing how mineral endowments and public policy factors affect exploration and development investment. Since 1998, the survey has expanded from just reviewing Canadian provinces to also include a number of states in the U.S. and a growing number of nations globally. In 2002, the survey was expanded to review the investment attractiveness of 45 jurisdictions including the Canadian provinces and territories (except Prince Edward Island), selected U.S. states (this year Alaska, Arizona, California, Colorado, Idaho, Minnesota, Montana, Nevada, New Mexico, South Dakota, Utah, Washington, Wisconsin, and Wyoming), Argentina, Australia, Bolivia, Brazil, Chile, China, Columbia, Ecuador, Ghana, India, Indonesia, Kazakhstan, Mexico, New Zealand, Papua New Guinea, Peru, Philippines, Russia, South Africa, Venezuela, and Zimbabwe.

Many in the mineral exploration community use this survey when considering investment risks and where best to spend exploration dollars regarding mineral resource development globally. Over the last 3-5 years, states in the U.S. have lost ranking in this survey. For instance, in 1998, Nevada ranked number 1 in the Policy Potential Index, number 4 in the Mineral Potential Index, and number 1 overall (Investment Attractiveness Index—IAI). By 2002, Nevada's ranking fell—to number 2 in the Policy Potential Index, to number 8 on the Mineral Potential Index, and to 4 overall. Table 1 compares the Investment Attractiveness Index for the 1998-99 Survey to that of 2002. In almost every case, the states lost ranking in the 2002, losing out to other Canadian provinces and countries around the world. Three states (ME, MI, MO) are not being considered in the 2002 survey.

Table 1.

State	1998-99 Survey Ranking	2002 Survey Ranking
AK	9	15
AZ	10	19
CA	25	38
CO	16	33
ID	11	32
ME	28	
MI	26	
MN	27	42
MO	21	21
MT	14	
NM	17	31
NV	1	4
SD	24	37
UT	12	27
WA	23	43
WI	30	44
WY	22	41
	of 31 Provinces, states and countries	of 44 Provinces, states and countries

Appendix A has figures and brief summaries from the Fraser Institute 1998-99 Survey. Figure 5 in Appendix A illustrates how the states ranked against one another. Appendix B has some of the figures from the 2002 survey, and Chart 1 illustrates how the states ranked against one another (from the 2002 survey data).

Some have suggested that the mineral potential of the U.S. has been tapped out, as stated by one Department of Interior official who suggested in 1999 that all of the gold in the U.S. had already been found (personal communication, 2003, John Dobra, Associate Professor of Economics, UNR). As an exploration geologist, I find this statement lacking reason or scientific basis. Several new discoveries in Nevada were made in 2001, well after this statement was made. These have mainly been at or near existing mine sites where most exploration dollars are being spent.

New discoveries are the result of focused scientific investigation, committed investment dollars, and ever changing and evolving technologies. These help to advance the understanding of geologic processes and mineral deposition, leading to mineral resource discoveries. Mineral deposits continue to be found in newly discovered, grassroots areas, as well as in "mature" geologic settings where mineral deposits have been previously discovered and defined. Advances in the geologic and exploration-related sciences during my career, have led me to revisit historic mining districts I had previously explored because they are "prospective" once again as a result of the advances in geologic sciences, data collection techniques and new technologies.

As an example, Nevada has been a world leader in gold production from surface open-pit mining operations since the 1970s. These are typically large, lower grade deposits, produced utilizing technologically advanced techniques. More recently with newer technologies and advances in the geologic understanding of mineralized areas, significant higher-grade underground deposits have been identified and developed—in areas where modern underground mining has not been the norm.

Despite these new discoveries, mining companies are investing less money in the U.S. and instead spending their research and development (exploration) dollars off shore.

I have personally experienced a drastic decrease in the funding of U.S.-based mineral exploration and development projects. My experiences are but a small window into the larger decreases seen in the exploration sector of the mineral development industry. Companies I have worked with since 1996 have chosen to cancel budgets here in the U.S. and re-channel their funds to "less politically risky places" such as Africa. That was a chilling statement, made to me by an investment client in 1997 in the face of the controversy surrounding the revisions of the 3809 Regulations and the Millsites Opinion—both of which created such an unstable investment climate that my client and others in the investment community decided to take their exploration and development dollars overseas. That exodus has not stopped since 1997.

TRENDS-IMPACTS

Regulatory and policy uncertainties initiated by the revisions of the 3809 Regulations and fueled by the Millsite Opinion, regulatory inconsistencies, bonding crisis, and access issues continues to deter the development of new mines in the U.S., with investment dollars being spent overseas at an ever increasing rate. The Fraser Institute 2002 survey reports that senior mining companies are now spending only 7% of their exploration budgets here in the U.S., while junior companies might be spending 10% on U.S. exploration projects. This is a radical decrease—seriously down from highs of roughly 50% of budgets being spent in the U.S. in the recent past (Figures 1 and 2).

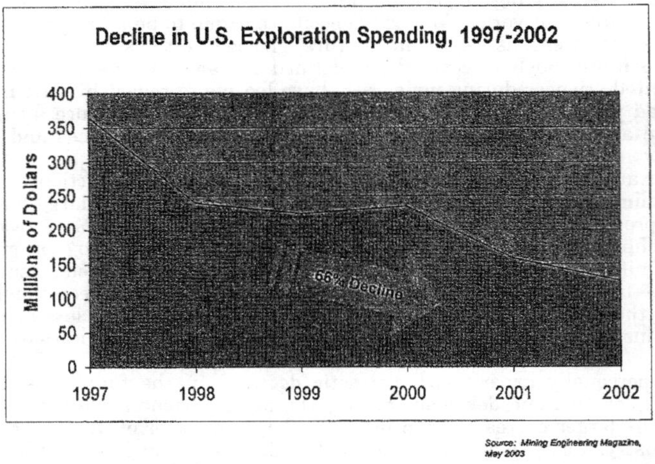

Figure 1.

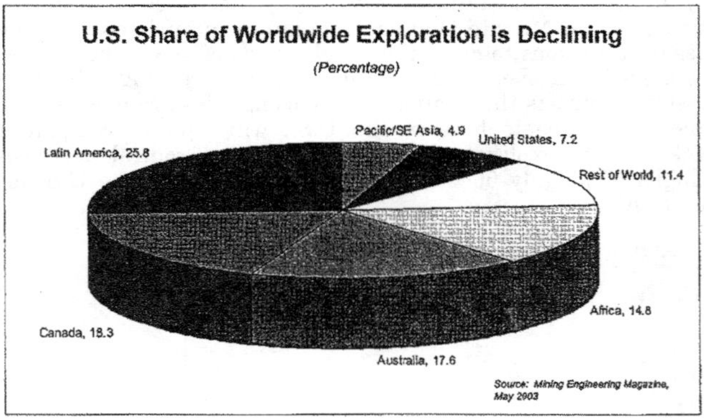

-- U.S. share dropped from 10% of total in 2000 to 7.2% in 2002, out of total worldwide budget of 1.73 billion.

-- U.S. exploration budget in 2002 was about 34% of the level it was in 1997.

Figure 2.

During the 1990's while the rest of the U.S. economy was booming, there was a serious decline in U.S. mining activity, a decline that continues today with some of the trends and impacts illustrated below:

- Since 1996, there has been a 73% decline in new claims.
- Exploration expenditures have continued to steadily decrease and grassroots exploration has virtually disappeared in the U.S. More money is consistently being spent on overseas projects.
- Mid-size producers and "juniors", generally where most exploration investment dollars come from, have chosen to invest overseas rather than in comparatively equal opportunities in the US.
- Large mining companies are replacing depleted domestic reserves by acquisition of producing properties through mergers rather than exploration for new prospects.
- Mining schools are being lost outright; some are being closed, and others are being consolidated and assimilated into other programs at universities—losing their mining focus and expertise.
- Greater challenges related to economic and national security issues.

The results—U.S. is exporting mining investment dollars and engineering talents and innovations to countries where metal mining is expanding at an ever increasing rate. As well, other losses include tax and related revenues and jobs at the local, state, and national levels; an increase in foreign reliance on foreign produced minerals and products; and other related negative economic impacts to rural American communities where mineral resources are generally, and more likely to be developed.

An Exploration Geologist's View

As an exploration geologist, I am particularly sensitive to this decrease in exploration funding here in the U.S.—not only when considering my own ability to make a living, but more importantly when considering the severe impact on rural communities due to this downturn. Mineral exploration activities are commonly focused in remote regions of the American west. Decreases in exploration spending here in the U.S. directly impact rural communities in these areas. Below are some of the very basic expenditures "in the day of the life" of an exploration geologist:

BASIC EXPENDITURES

Hotels – approximate	$60/day
Meals – approximate	$35/day
Fuel, based on 100 mile roundtrip @$0.45/mi	$45/day
Total for these basic expenses	**$140/day**

Taking this a step further, in a good year I might work in the field approximately 200 days, living and working away from my home and staying in rural communities throughout the west. That equates to approximately $28,000 in expenditures funneled into rural communities. If 100 geologists were working regularly to explore the nation's mineral resources, this number might jump to roughly $2.8Million dollars. These are dollars that would likely be spent in rural communities if exploration and development were being encouraged here in the U.S.—based on just 100 geologists working. This is a very conservative estimate of exploration spending.

The calculations above only include the most basic of expenditures, and do not begin to summarize the truly large expenses related to additional exploration investments—rock sampling, assaying, drilling, engineering evaluation, metallurgical testing, to name but a few—and are further examples of jobs and revenue possibilities in a strong exploration climate, for many communities across the west.

IMPEDIMENTS TO DEVELOPING MINERAL RESOURCES

"Mining is difficult no matter where you go in the world. But at least in other countries I know that if I meet the regulatory and legal requirements, I'll get a permit ... I don't know that in the U.S.—I never know if or when I'll finally get a permit—even if I can demonstrate the mine will be in full compliance." Senior Mining Executive, 2003

I have been witness to a declining mineral development business here in the U.S., while watching and participating in a relatively stable minerals exploration and development industry overseas. I have personally experienced radical decreases in funding and eliminations of budgets for U.S.-based exploration programs due to the "uncertainties" that are associated with U.S. laws, regulations, and policies regarding mineral development. Most companies willing to invest in mineral properties worldwide regard the U.S. as highly prospective for mineral discovery, but highly risky regarding regulatory processes and policies, with an increasingly cumbersome and negative permitting regime.

The perceptions of "uncertainty" are and continue to be aggravated by many factors. I will discuss five, including: the revisions to the "3809" regulations; the release of the former Solicitor's Millsite Opinion; inconsistent interpretation and implementation of existing regulations; the bonding crisis, and access issues.

Negative Impact of the 3809 Revisions

The Bureau of Land Management's October 2001 revisions to its Section 3809 regulations were necessary to achieve consistency with the recommendations of the National Academy of Sciences (NAS) in its report, "Hardrock Mining on Federal Lands," completed in 1999. The NAS study, requested by the governors of the Western States and mandated by Congress, established a clear and scientifically based benchmark for appropriate environmental protections associated with hardrock mining on federal lands. Importantly, the revised 3809 regulations include improved bonding provisions, ensuring that adequate funds be guaranteed for reclamation of mining operations, a change that was supported by the mining industry. One of the key findings by the NAS team was that the existing regulations were generally adequate in providing environmental protection. The NAS team also indicated the greatest improvements that could be realized regarding the 3809 regulations would be if there were more consistent interpretation and implementation in the field, as well as better agency management and staff training. These recommendations have yet to be implemented in the field.

Although the U.S. mineral industry was supportive of updating the 3809 regulations and worked with DOI and others to achieve this, the revision process initiated a climate of uncertainty regarding regulations. This was then followed by the 1997 Millsite Opinion, which fueled the exodus of mining-related investment dollars from here in the U.S. to properties overseas, which continues today.

Negative Impact of the Millsite Opinion on U.S. Mining

The General Mining Law authorizes staking mining claims on public lands for the purpose of exploring for and developing "locatable" mineral deposits, including base metals (copper, lead, and zinc); precious metals (gold, silver, and platinum group metals); uranium; and certain industrial minerals including gypsum, lithium, borates, barite, diatomite, and some clays and limestones. This law also defines several different types of claims including lode, placer, millsites, and tunnel sites that are used for different applications in specific situations.

In 1997, the Department of the Interior (DOI) Solicitor issued an opinion on the use of millsite claims, applying a maximum allowable ratio of one-to-one between lode claims (approximate 20-acre claim staked on valuable mineralized land), and millsite claims (5-acre claim staked on non-mineralized ground, to be used for mining facilities and infrastructure). The Millsite Opinion is wrong and has no basis in law or policy; this arbitrary ratio is a radical departure from the way in which the Department of Interior interpreted and administered the 1872 Mining Law since its inception. In 125 years of judicial interpretation, not one case has addressed or discussed or implied a ratio between lode claims and millsite claims. Furthermore, the opinion is expressly contrary to long-standing BLM and USFS policy:

- 1991 BLM Manual at Section 3864.1.B provides "A millsite cannot exceed 5 acres in size. There is no limit to the number of millsites that can be held by a single claimant."
- 1990 USFS Manual at Section 2811.33 provides "The number of millsites that may legally be located is based specifically on the need for mining or milling purposes, irrespective of the types or numbers of mining claims involved."
- The California State BLM office has records indicating that multiple millsites have been the practice since at least as far back as 1903.

Nothing in the 1872 Mining Law suggests a one-to-one millsite-to-lode claim ratio. Rather, the criteria used by federal land managers to evaluate the appropriate use of millsite claims was that the land should not be mineralized and there should be a demonstrated need for the land upon which the processing and ancillary facilities were built. The Millsite Opinion is a back-door administrative attempt to change the U.S. mining law to remove the existing right to use as much of the surface of non-mineralized public land (millsite claim) as is reasonably needed to support the development of a mineralized claim (lode or placer)

An Example—Exploration Stopped

The 1997 Millsite Opinion has helped to fuel the perception of regulatory uncertainty here in the U.S., contributing to shifting mining investments overseas. I personally had $2 Million U.S. pulled from a proposed U.S. exploration budget and channeled to Africa—because the exploration company perceived that it would be less "risky", from a regulatory perspective, to work in Africa than here in the U.S. This occurred in 1998, and was the direct result of the unease caused first by the revision process of the 3809 Regulations, and seriously exacerbated by the 1997 Millsite Opinion.

An Example—Mine Development Stalled

According to Greg Hahn, President and CEO of Summo USA Corporation, the company's initial goals prior to 1999 were the development of copper resources in the U.S. that were too small for the major copper producers. Their current focus has shifted to copper projects primarily outside of the US. This shift is a direct result of the adverse investment impacts created by the former Solicitor's Millsite Opinion. As well, the uneven leverage afforded anti-mining groups in opposing and appealing projects, and the uncertainties in the regulatory and permitting arenas creates a negative investment climate here in the US, further prompting this company to seek investments offshore. The table below in part illustrates this:

Summo USA Corporation – Exploration and Development Expenditures

Years	US Projects	Foreign Projects
1995-1999	$14Million	<$1Million
1999-Present	$<2Million	>$9Million

Inconsistent Interpretation & Implementation of Existing Regulations

The USA needs a dose of "environmental realism based on good scientific/ engineering policies." Evaluations Manager, senior mining company (Fraser Institute 2002 Survey)

Mining is a modern, high-tech, environmentally responsible industry providing minerals essential to the nation's economic growth, to its national security and to American's quality of life. Laws and regulations governing mining should provide clear and consistent environmental guidelines, facilitating compliance efforts by exploration and mine operators. This would begin to reduce regulatory uncertainties, helping to attract needed capital to the domestic mining industry.

I have first hand experience managing permitting requirements on exploration projects (Notices of Intent and Environmental Assessments). As well, I was the company lead in 1997 on a team completing a detailed Environmental Impact Statement (EIS) through the Bureau of Land Management on a proposed mining project within 50 miles of Reno, Nevada. This EIS was completed in less than 18 months on a complex mine proposal.

Through my various permitting experiences, and comparing notes with my peers addressing similar permitting concerns, the mounting inconsistencies surrounding interpretation and implementation of existing laws and regulations creates a major hurdle in mineral development here in the U.S. This translates to serious permitting delays and related elevated costs to projects. One result is a lengthened permitting process, with an EIS taking at least by 2-4 times longer to complete now than it did 5 years ago. This does not factor in the possible lengthy delays brought on by lawsuits—which has become a "next step" by groups opposing mining here in the U.S.

Other areas of uncertainties in the permitting process include:

- Uncertainty and legal confusion over recent developments involving Native American sites
- Endangered Species Act
- Uncertainty regarding possible new Mining Law legislation
- Uncertainty of legal appeals
- Transparency issues—agencies not operating to the same transparent standards as the industry is required to, such as was seen at the Crandon Mine Project, WI ("Under the Guise of Environmental Protection" EPA Revealed, National Wilderness Institute report, 2000).

Bonding—Predicaments & Impediments

Some form of financial assurance is a prerequisite to obtaining permits, and this has traditionally been in the form of a surety bond. Mining companies, both large and small, are experiencing increased difficulty in securing the necessary bonds to satisfy financial assurance requirements under various regulatory programs. Today, surety bonds for mining related obligations are virtually unavailable at any price, with or without collateral. Below are some reasons for this downturn:

- The Enron, K-Mart, Global Crossing and W.R. Grace bankruptcies
- September 11, 2001 terrorist activities (insurance industry lost about ° of its $150 billion pre 9-11 capital)
- The surety industry is experiencing increasing losses on non-mining obligations
- Mine reclamation bonds represent less than 1% of the surety business line, but have the longest tails

Regulatory impediments have contributed to the surety industry's decision to place its capital in businesses other than mining, ones with more favorable risk/reward profiles. Such impediments include:

- Large bond amounts, inflated due to excessive contingencies, speculative assumptions, and other cost factors (3d party, Davis Bacon, excessive overhead, etc.). All of this results in an increase of approximately 40% above the actual cost of reclamation. This includes exploration projects as well.
- Glacial pace at which BLM (at least in NV) reviews and approves bonds
- Severe reluctance to release bonds once the reclamation work has been completed
- Complex and constantly changing regulatory schemes

The mineral development industry needs assistance in addressing bonding impediments toward finding a workable solution.

Access Issues

According to the GAO, as of September 30, 1993, the federal government owned approximately 650 million acres and was managing 271 million acres (43.7%) for conservation purposes. In the ensuing ten-year period additional lands have been acquired and are designated as managed for conservation purposes through various administrative and legislative processes. Some of these designations are listed below:

- California Desert Protection Act—7.7 million acres
- Clinton's National Monuments—approximately 4 million acres plus
- Previous Administration's Roadless Rule—60 million acres

These designations increased the total federal acreage managed for conservation purposes from approximately 44% in 1993 to 55% at present. These numbers do not include endangered species habitat.

Lands managed for conservation purposes and military reservations are generally not open to mineral entry. Additional land designations and programs that impact access and impede mineral exploration and development include:

- Time restrictions on physical exploration of a prospect that involves building access roads, drilling or trenching, in order to accommodate the mating, early life stages, feeding and watering or migrating habits of threatened and endangered species. In Eastern Nevada there are numerous examples where exploration drilling of prospects adjacent to operating mines was severally restricted for these purposes.
- Withdrawal of areas prospective for mineral discovery from mineral entry, such as Crown Butte (defined and designated mineral reserves) and the Sweet Grass Hills (existing claims and ongoing mineral exploration projects) in Montana
- Indian Sacred Sites (one of the reasons given for setting aside the Sweet Grass Hills and the denial of Glamis Gold's permit in California)
- Land exchanges
- Having a prospect or discovery in close proximity to an area that has been set aside for conservation purposes, even if it has been expressly left out of the conservation area because the area is prospective for or is known to contain valuable mineral deposits (Crown Butte - Montana)
- Wilderness Study or Roadless Areas (RARE I & II Lands, which are now incorporated in the Roadless Rule)

Many of these withdrawals and designations occurred without review for: mineral potential; renewable and/ or non-renewable energy potential; and impacts to existing communities. In some cases, known mineral and energy resources were dismissed—as was the case with the Grand Escalante-Staircase Monument listing. These numbers are conservative and do not include other actions by the previous Administration, such as former Secretary Babbitt's removal from mineral entry of almost 3 million acres. This occurred in the first 9 months of 1999, and includes twenty-year moratoriums for mineral entry over areas—Crown Butte and Sweet Grass Hills in Montana—with known, defined and engineered mineral resources.

In contrast to the lands set aside for conservation purposes, mining in the U.S. has impacted approximately 6 million acres slightly more than two tenths of a percent. About 45 percent of the areas impacted by mining have been reclaimed and many other areas are still actively being mined.

IN SUMMARY

The U.S. needs to "Get back to the reality that the U.S. is dependent on metals to make the economy grow and prosper—same with energy." President, junior mining company (Fraser Institute 2002 Survey)

The permitting and regulatory processes have become slower and more litigious with each passing year. We have seen many viable projects taken into the courts after interminable permitting reviews, studies, and processes. As a result, the investment community is taking its monies overseas; we are losing jobs, revenues, and income to this flight overseas; and we are becoming more reliant on foreign sources for the minerals that we consume daily.

For me, the displacement of the domestic mining industry raises a poignant dilemma.

In all of the places overseas I have had the opportunity and pleasure to work, one recurring question is always asked of me by the locals—"Why are you here taking my minerals?" Is this a question of mineral or foreign policy, or a combination of both? Most of the time, the question is rhetorical or they will answer it with this assessment—in the U.S., we are not allowed to access and permit our own resources, so we are forced to travel to other countries to find the materials that we need to feed our consumption.

If we are to reverse the current downward trends in the domestic mining industry, maintain our leadership role in the development of mining technology, environmental practices and enhance our market share of this crucial industry, we will have to develop a comprehensive Domestic Minerals Policy.

I believe one of the factors that have hampered the legislative process in the development of a strong Domestic Minerals and Energy Policy is the perception that Americans are opposed to mining in the U.S. and believe that federally managed lands should be set aside for purposes other than resource development. Survey research does not support that perception. According to a nationwide survey of 800 registered voters conducted by Market Strategies for the National Mining Association last year,

- 90% believe we need "a National Minerals Strategy to ensure our quality of life in the future";
- 73% say lands owned by the United States should be open to mining, provided the land is reclaimed [as required by law];
- Only 22% say these lands "belong to the public and should be set aside for future generations to enjoy and should not be used for mining, forestry or ranching."

Mining is a difficult venture no matter where the project is located around the globe. The modern mining industry must address many issues and concerns while developing projects—social-cultural considerations, engineering requirements, possible environmental impacts, economic needs, and many other concerns. Evaluations are completed within detailed and lengthy studies and communications. Investors and the people developing mineral projects should be able to operate with some level of confidence—if the regulatory requirements and laws are met, then mineral resource development can follow. That is a confidence realized more often on overseas projects. There are just too many inconsistencies in implementing the established regulations here in the U.S. to achieve that same level of confidence. The protracted, uncertain and contentious permitting processes here in the U.S.—for all aspects of mineral development, from exploration through production—creates an excessively uncertain investment atmosphere and has lead to a diversion of exploration funds to countries with more streamlined, transparent and expedited permitting processes. More and more investors view other countries more positively than the U.S., where monies can be put to work to benefit local and national economies.

Solutions

Many in the mining community believe that the inconsistencies and uncertainties related to permitting mine projects can be addressed. Below are suggestions toward correcting some of these inconsistencies:

- Provide firm time guidelines and deadlines—for both the information gathering and review processes.

- Review the adequacy of BLM and Forest Service staff and resources devoted to regulating mineral exploration and mining operations.
- Update technical and policy guidance documents on a regular basis.
- Increase and improve agency and stakeholder participation in the NEPA process from its earliest stages.
- Expedite the review of permit applications for exploration projects affecting fewer than 5 acres of Forest Service-managed lands.
- Require financial assurance for all mining and exploration activities that are not classified as casual use.
- Mandate Plans of Operation for any mining or milling operation regardless of size.
- Develop criteria and procedures for modifying Plans of Operation.
- Adopt regulations that define temporary closure and require interim management plans.
- Plan for and assure long-term, post-closure management of closed and reclaimed mines.
- Provide authority to issue administrative penalties and develop clear guidelines for involving other state or federal enforcement authorities.
- Modify existing environmental laws and regulations to allow and promote industry cleanup of abandoned mines and remove institutional and legal barriers currently thwarting such cleanup.
- Secure Congressional funding for aggressive and coordinated research programs on the environmental impacts of hardrock mines.
- Require the losing party to pay all costs and attorney fees if they challenge agency decisions in court.

As one mineral economist concluded recently—the most serious threat to the mining industry's long term sustainability in the U.S. is the regulatory changes made in the final four years of the Clinton administration—including revisions to the 3809 Regulations and the Millsite Opinion. Let's work together to reverse these impediments and turn around the current trend towards offshore investment and greater reliance on foreign mineral sources. Let's work to develop and implement a working National Minerals Policy that serves to provide national and economic security now and into the future.

Chart 1.

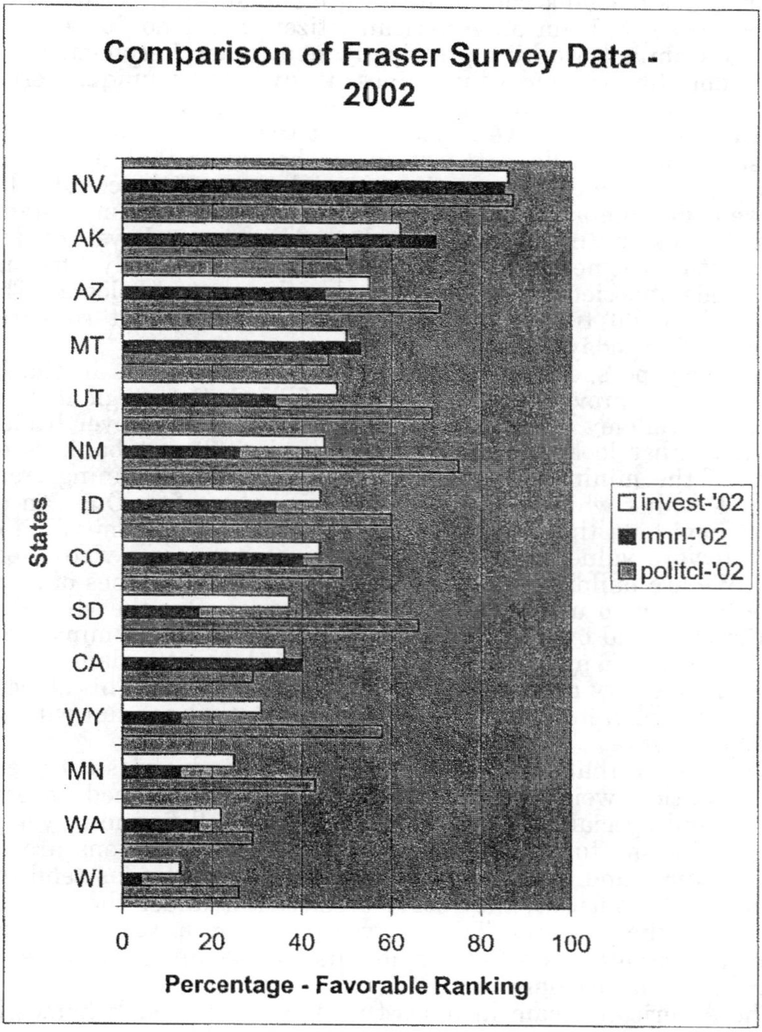

politcl-'02 – Political Potential Index, 2002 Fraser Survey
mnrl'02 – Mineral Potential Index, 2002 Fraser Survey
invest'02- Investment Attractiveness, 2002 Fraser Survey

States included in the 1998-99 Survey, but excluded in the 2002 Survey – Maine, Michigan and Missouri.

NOTE: Appendices A and B have been retained in the Committee's official files.

Mrs. CUBIN. Thank you, Ms. Carpenter.
The Chairman now recognizes Douglas B. Silver.

STATEMENT OF DOUGLAS B. SILVER, PRESIDENT, BALFOUR HOLDINGS, INC.

Mr. SILVER. Thank you for inviting me to speak here today. I too am a geologist. I am an American citizen and I do 90 percent of my work abroad, working principally as a mineral economist, helping people finance and value mines. It gives me a unique perspective because I do not work too much in the United States.

Others before me have spoken about the supply/demand picture for strategic minerals, and you should be aware that this list has changed over the century as the material needs of the country have evolved. For instance, the early list focused on wartime commodities, but the nature of war has changed as new alloys are developed with each new generation of armament. Similarly, the material needs of society have also shifted with new technologies. These changes should trigger an important dialog on what constitutes "strategic" in today's world.

From my perspective, however, I see several trends in the U.S. economy that provide guidance for the future of strategic minerals. I see an economy where service industries dominate over basic industries. Just look at the profits of the investment bankers over those of the mining industry. But unlike banking, mining creates tangible value while investing creates paper profits. One can only guess how long the world's money can shift away from value creation toward value harvesting. The new economy is a virtual economy. We are building a virtual Nation where the engines of growth have little or no underpinning in hard assets. But the virtual Nation is premised on a false reality. Today's economy reminds me of an upside down pyramid where wealth and its attendant luxuries depend on a very narrow base of true value creation. This structure creates an inherently unstable platform, one that can fall with very little assistance.

We also see that tourism and recreational industries carry enormous political weight in Washington. This has spawned an era of NGO's and special interest groups that are well financed, yet provide no income to the U.S. economy. Their contributions are both philosophical and intangible, but because their agendas tend to be for a specific purpose, they serve a select few under the pretenses of serving the majority. Their power is immense, yet they pay no taxes and create few jobs. Despite their noble intentions, they are a drain on the economy.

The American dream includes the right to own one's home. This hope has seen massive expansions in populations and their special needs. In my home State of Colorado, I see many middle Americans also purchasing second homes in the mountains. But where does the land come from for these new residences? It comes from the agricultural lands and other rural areas, resulting in less land being available for basic industry. We also see more and more lands being set aside for recreational users, lands whose commercial contributions to the tax base and job creation are being limited by their restricted use.

Special interest groups often talk about ecosystems and rightly so. The reduction in available land for mineral exploration and development is reaching the point where our industrial ecosystem is severely threatened. It seems that every time a new deposit is discovered in some remote region, that area instantly becomes a beehive of sacred sites, premier vistas and unique habitats. What most do not realize is that the geologic processes responsible for building mountain ranges are also responsible for developing ore deposits. So as each new vista view is set aside it is preferentially depleting our natural resource base.

So I ask you to think about the following issue. Today the world's largest mineral companies tend to be public companies, so why is the United States the wealthiest Nation in the world and the largest consumer of metals, yet is the home to so few mineral companies? Based on research conducted by my firm, U.S. domiciled companies only represent 11 percent of the world's public mineral companies. Approximately one half of the U.S. companies are active, but only about 20 are focused on hard rock minerals, the balance being focused on coal and industrial minerals. Wouldn't you think the largest consumer of metals would also be the largest supplier of metals?

The proof is in the fact that Newmont Mining, the world's largest gold producer and an American company, derives 57 percent of its annual production from overseas. Ever wonder why? We found no evidence that the lands had been thoroughly explored, but there is plenty of evidence that there is less land available to explore and that the difficulty in conducting even simple exploration on Federal lands serves as a strong incentive to work elsewhere.

The proof is in the fact that the mineral investment in the State of Montana is so bad that only the Russians are willing to take on the risk.

The proof lies in the Government's abandonment of the U.S. Bureau of Mines, the severe budget cuts to the U.S. Geologic Survey and the decisions you are currently making that are compromised by the lack of information you are currently receiving. Ask the Mineral Information Team at the USGS how many positions they have vacant and why they have no budget to fill these slots. Take the time to understand that these positions were identified as important, yet those responsible for slashing their budgets felt the information will have no material impact on your ability to make smart decisions.

By contrast, Canada represents 58 percent of the known mineral companies including the vast majority of exploration firms. Why has Canada become the leading home for public mineral companies? The answer lies in Canada's commitment to developing its natural resources. I believe the Canadian Government strives to create solutions that permit the cohabitation of special interest groups and the engines of production, while ours avoids confrontation with the NGO's at the expense of commerce. Unlike our Securities and Exchange Commission, the Canadians utilize global reporting standards. Unlike the EPA, the Canadians recognize the legal rights of working people to use their lands for commercial purposes. Unlike the United States, the Canadian insurance industry seems to work hard at providing bonding for new mine development because it understands the importance of commerce to the national economy.

This brings us back to the issue of strategic minerals. In my opinion, all minerals are strategic, especially when they are no longer available to society. Strategic is what we call phosphate when 99 percent of our imports come from Morocco. Yttrium, an element used in television screens and magnets, is strategic because more than one half of our imports come from China. We import almost 70 percent of our manganese from Gabon and South Africa. Any mineral or metal that requires substantial import should be considered strategic. This means that all metals and minerals are strategic.

I would like to leave you with two concepts to think about.

What do we as a Nation want to do with our minerals? Will we always have the economic and military might to fulfill our natural resource needs by any means we deem necessary? If we believe we will, then we should continue along the current and past administration's policies of ignoring the domestic mineral industry. I can only surmise that Washington's lack of progress reflects your comfort levels with these existing policies. But like a drug addict, the longer you ignore the problems, the harder it will be to kick the bad habit. From my observations, the difficulties of exploring or developing a hard rock mine in this country are immense. They will only get worse if you continue along your current path.

If you believe that the strength of a Nation is founded on the abundance and employment of its natural resources and that basic industries play a critical role in building lasting value for the national economy, then we need a dramatic shift in the Federal Government's attitude about U.S. minerals and how we intend to manage them.

We need a Government that is proactive in its support of mining. Action items I would like to see include the courts quickly identifying frivolous lawsuits filed by the elitist and permit companies to seek monetary damages if the case is deemed to be frivolous. A typical U.S. mine takes 10 years or longer from discovery to production, whereas elsewhere in the world this discovery will be placed in production in four or 5 years. Discovery is typically made in the first year. Two or 3 years are then spent in conducting follow-up work and completing feasibility and determining the economic viability of the project. Most mines take 1 year—

Mrs. CUBIN. Could you sum up your statement?

Mr. SILVER. Ma'am, there are a lot of issues in the mining industry that need to be addressed, and we see the inactivity of the Federal Government as being the biggest single problem we have. We don't care which way you go, but we really wish you would pick a direction so we can get on with our business.

Thank you.

[The prepared statement of Mr. Silver follows:]

Statement of Douglas B. Silver, President, Balfour Holdings, Inc.

Thank you for inviting me to speak today. Others before me have spoken about the supply-demand picture for strategic minerals. This list has changed over the past century as the material needs of the Country have evolved. For instance, the early lists focused on war-time commodities, but the nature of war has changed as new alloys are developed with each new generation of armament. Similarly, the material needs of Society have also shifted with new technologies. These changes should trigger an important dialogue on what constitutes "Strategic" in today's modern world.

From my perspective, I see several trends in the U.S. economy that provide guidance for the future of Strategic Minerals. I see an economy where service industries dominate over basic industries. Just look at the profits of the investment bankers over those of the mining industry. But unlike banking, Mining creates tangible value, while investing creates paper profits. One can only guess how long the world's money can shift away from value creation toward value harvesting. The New Economy is a virtual economy. We are building a Virtual Nation where engines of growth have little or no underpinning in hard assets. But the Virtual Nation is premised on a false reality. Today's economy is starting to remind me of an upside pyramid where the wealth and its attendant luxuries depend on a very narrow base of true value creation. This structure creates an inherently unstable platform; one that can fall with very little assistance.

We also see that tourism and recreational industries carry enormous political weight in Washington. This has spawned an era of NGOs and special interest groups that are well financed yet provide no income to the U.S. economy. Their contributions are both philosophical and intangible. But because their agendas tend to be for a specific purpose, they serve a select few under the pretenses of serving the majority. Their power is immense, yet they pay no taxes and create few jobs. Despite their noble intentions, they are a drain on the economy.

The American dream includes the right to own one's home. This hope has seen massive expansion in populations and their special needs. In my home state of Colorado, I see many middle Americans also purchasing second homes in the mountains. But where does the land come from for these new residences? It comes from agricultural lands and other rural areas, resulting in less land being available for basic industry. We also see more and more lands also being set aside for recreational users: lands whose commercial contributions to the tax base and job creation are being limited by their restricted uses.

Special Interest Groups often talk about ecosystems and rightly so. The reduction in available land for mineral exploration and development is reaching the point where the Industrial Ecosystem is severely threatened. It seems that every time a new deposit is discovered in some remote region, that area instantly becomes a beehive of sacred sites, premier vistas and unique habitats. What most do not realize is that the geological processes responsible for building mountain ranges are also responsible for developing ore deposits. So as each vista view is set aside, it is preferentially depleting our natural resources base.

So I ask you to think about the following issues. Today, the world's largest mineral companies tend to be public companies. So, why is the United States the wealthiest nation in the world and the largest consumer of metals, yet it is the home to so few mineral companies? Based on research conducted by my firm, U.S.-domiciled companies represent only 11% of the world's public mineral companies. Approximately one-half of these U.S. companies are active but only about 20 are focused on hard rock minerals (the balance being principally focused on coal and industrial minerals). Wouldn't you think that the largest consumer of metals would also be the largest supplier of metals?

The proof is in the fact that Newmont Mining, the world's largest gold producer and an American company, derives 57% of its annual production from overseas. Ever wonder why? We have found no evidence that the lands have been thoroughly explored. But there is plenty of evidence that there is less land available to explore and the difficulty to conduct even simple exploration on Federal lands serves as a strong incentive to work overseas.

The proof is in the fact that mineral investment in the State of Montana is so bad that only the Russians are willing to take on the risk.

The proof is in the fact that the number of staked claims in Nevada dropped precipitously when the annual rents were doubled.

The proof lies in the Government's abandonment of the U.S. Bureau of Mines, the severe budget cuts to the U.S. Geological Society and the decisions you are currently making that are compromised by the lack of information you are currently receiving. Ask the Minerals Information Team at the USGS how many positions they have vacant and why they have no budget to fill these slots. Take the time to understand that these positions were identified as important yet those responsible for slashing their budgets felt the information will have no material impact on your abilities to make smart decisions.

By contrast, Canada represents 58% of the known public mineral companies, including the vast majority of exploration firms. Why has Canada become the leading home for public mineral companies? The answer lies in Canada's commitment to developing its natural resources. I believe the Canadian government strives to create solutions that permit the cohabitation of special interest groups and the engines of production while ours avoids confrontation with the NGO's at the expense of commerce. Unlike our Securities and Exchange Commission, the Canadians utilize global reporting standards. Unlike the EPA, the Canadians recognize the legal rights of working people to use their lands for commercial purposes. Unlike the United States, the Canadian insurance industry seems to work hard at providing bonding for new mine development because it understands the importance of commerce to the national economy.

This brings us back to the issue of strategic minerals. In my opinion, all minerals are strategic, especially when they are no longer available to Society. Strategic is what we call phosphate when 99% of its imports come from Morocco. Yttrium, an element used in television screens and magnets, is strategic because more than one-half of our imports come from China. We import almost 70% of our Manganese from Gabon and South Africa. Any metal or mineral where we require substantial imports should be considered strategic. Any metal that provides underpinning value for our Virtual Nation should be considered strategic. This means that all minerals and metals are strategic.

I would like to leave you with two concepts to think about.

1. What do we, as a nation, want to do with our minerals? Will we always have the economic and military might to fulfill our natural resource needs by any means we deem necessary? If we believe we will, then we should continue along the current and past Administrations' policy of ignoring the domestic minerals industry. I can only surmise that Washington's lack of progress reflects your comfort levels with these existing policies. But like a drug addict, the longer you ignore the problems, the harder it will be to kick the bad habit. From my observations, the difficulties of exploring or developing a hard-rock mine in this country are immense and will only get worse if you continue along your current path.
2. If we believe that the strength of a nation is founded on the abundance and employment of its natural resources and that basic industry plays a critical role in building lasting value for the national economy, then we need a dramatic shift in the Federal government's attitude about U.S. minerals and how we intend to manage them.
   - A. We need a government that is proactive in its support of mining. Action items I would like to see include having the courts quickly identify frivolous lawsuits filed by the elitists and permit companies to seek monetary damages if the case is deemed to be frivolous. A typical U.S. mine takes ten years or longer from discovery to production whereas elsewhere in the world this discovery could be placed into production in four or five years.. Discovery is typically made in the first year of exploration. Two to three years are then spent conducting follow-up work and completing feasibility studies that determine the economic viability of the project. Most mines take approximately one year to construct. This suggests that five years are spent fighting for our rights in the courts. Hardly seems fare to me. Can you imagine how Congress would function if we only approved your budget once every ten years? The courts need stronger measures so that everyone's rights are protected, regardless of their perceived political correctness.
   - B. We need an SEC that adopts global standards rather than develops its own eclectic sets of rules. I understand that its intention is to protect the shareholders. But its morass of unwritten regulations and conditional approvals are preventing management from fulfilling their fiduciary duties.
   - C. We talk about preserving the future for our grandchildren, but do little or nothing to preserve and transfer the technical expertise of our aging talent pool. I am 48 and am considered one of the young ones in our business.
   - D. We do little to insure that our mining schools survive. There is an argument that we need less mining education because there is less mining in our country. But what we do not realize that there is less mining because there are too many impediments to building a mine. Consequently, students choose other majors forcing the schools to close or cut back. This results in less education and cutting-edge research that could provide solutions to today's problems.
   - E. What worries me the most is the government's lethargy. It was quite evident that Clinton's administration was against Mining but I must confess that I have not observed any contribution by President Bush's people either. Meanwhile our existing mines are being depleted which increases the nation's need for foreign supplies in the same way we have strong foreign dependence on oil. This worries me.

Now I know that your policies reflect the will of the people, but you also need to take a leadership role in protecting their interests by insuring there are inexpensive supplies of future metals. The same supplies that keep the price of cars affordable, that protect our high standards of living and keep our rural communities prosperous. We miners stand with pride as our forty-year old copper mines continue to operate, but we have grave concerns that there is little new supply being sought to replace them. Making money in mining is a tough business, particularly when the industry has to compete so ferociously at the global level. But when the U.S. government inhibits progress through its inactivity, and makes no effort to support its basic industries, global forces will draw this talent pool and capital expenditures to those regions where it is wanted. So I implore you to let us know our fate.

Thank you.

Mrs. CUBIN. Thank you.

I would like to start the questioning with Ms. Carpenter. If we improved our policies toward availability of Federal lands for minerals development, do you believe that we can find significant amounts of specialty minerals within our borders to reduce our reliance on foreign supply?

Ms. CARPENTER. Absolutely, Madam Chairwoman. The advances in understanding of geologic processes, the advances in technologies provide us with opportunities that are infinite. I've traveled across the West looking at mineral deposits, and I continue to go back to zones or in areas that I've looked at previously because the science has changed, the technologies have changed. It's an ever-dynamic process. And to say that once you have walked across a piece of ground and you can make an assessment that there is nothing there is wrong.

Mrs. CUBIN. In your testimony, you discussed a survey of American attitudes toward mining on public lands. What were the results of that survey?

Ms. CARPENTER. In that survey, it was a survey conducted by Market Strategies for National Mining Association. In that survey some of the summaries were 90 percent of the people surveyed believe that we need a national mineral strategy to ensure our quality of life in the future. 73 percent say lands owned by the United States should be open to mining, providing the land is reclaimed as required by law. And only 22 percent say that these lands belong to the public and should be set aside for future generations to enjoy.

Mrs. CUBIN. Could you tell me, just list a few things that you think ought to be included in the national mineral policy that you recommend.

Ms. CARPENTER. In my testimony, starting on page 14 and going to 15, I think there are 14 suggested recommendations, suggested changes. Some of them are: Provide firm time guidelines and deadlines within the permitting process and within the regulations; review the adequacy of BLM and Forest Service staff and resources devoted to regulating mineral exploration and mining operations; expedite the review of permit applications for exploration projects; secure congressional funding for aggressive and coordinated research programs on the environmental impacts of hard rock mines. It is a broad spectrum. And finally, require the losing party to pay all costs and attorney fees if they challenge agency decisions. We find that to be an incredible impediment.

Mrs. CUBIN. I sympathize with that and agree that it would be and is a terrible impediment. Unfortunately, we try to do that with all parts of public lands, whether it is harvesting timber, whether it is bringing up coal, whether it is grazing. Whatever activity it is on the public lands, frivolous lawsuits seems to be the method that the opposition has chosen to keep people from using the public lands. And it is a very frustrating situation, although I agree with you very much, and I think we should continue to try to make that happen.

Mr. Hanes, how will our downward trend for minerals production and manufacturing in general affect our economic future, and especially, how can we rebound from the situation that hurts small business interests virtually in every region of the country?

Mr. HANES. That is a profound question. I am not sure I can even treat it adequately. But if you look at the sector that I am most familiar with, that is what we have classified as strategic metals sector, as my colleague Mr. Noel described, we are under economic pressures from a lot of directions. Some of that is simply economic pressures. We are dependent on the aircraft industry, for example, in general, both military and commercial. And you see a lot of that decline in particular in the post-9/11 period.

But the way that we can reverse this trend, in my opinion, is to pay, you know, more attention to those applications of these strategic and critical metals that are, frankly, hidden in these systems. I mean, you know, people are just simply unaware. And I talk about that a little bit in my testimony, the criticality of these small components that make these very sophisticated systems work. And in order to do that, you need to sustain the productive capability here in this country. And it can be—if indeed, in a lot of those critical applications, for example, we don't have that manufacturing capability, it can be in fact a weapon of economic warfare.

I mean, there are some examples that are talked about and critical components that go into tactical missile systems, for example, that are generated in China that suddenly became a little bit—and this is sort of—and I am not sure I have factual information here, but you hear stories where these critical components, all of a sudden during the Iraq campaign, became a little less available. And so people were unable to upgrade systems and supply systems that were required. There are just any number of those kinds of components that need to be sustained.

Bottom line is that, you know, protectionism probably isn't the answer. It is, really, assuring that we have the strong manufacturing and technological base to support that sophisticated manufacturing here in this country, to assure that it is available in the future.

Mrs. CUBIN. Thank you. Mr. Silver, you mentioned in your testimony that Canada has a commitment to developing its natural resources. About 10 to 15 years ago, a number of the mining provinces were headed down the same predicament that we are currently facing. What turned things around in Canada? Do you know?

Mr. SILVER. Yes, I think the thing that you'd notice the most difference is at the Federal level. The Canadian Federal Government seems to go out of its way to get people to work toward a solution quickly. A good example is this brand-new nickel deposit that has been found in Labrador called Voisey's Bay. A lot of groups were against it being constructed. The government got everybody to sit down, and in a period of two or 3 years they came to a solution. Now the project is moving ahead. We don't see that from the U.S. Government. So I think it is the proactive role that the Federal Government takes in getting people together and pushing them toward finding an answer.

Mrs. CUBIN. It seems to take forever for the Federal Government to get acting. We have known for years that we were going to have a natural gas shortage and that there would be a crisis. And just to get the problem to the attention of the American people and the media sometimes seems overwhelming. But I certainly think that we need to begin, and hopefully this hearing will be a beginning, for a national mineral policy or to at least expose the need for one. Because frankly, I haven't heard much talk about that in Washington. I think there a few people, mostly on this Committee, that are aware of it, but otherwise I think there is just really a void in information in this regard.

Mr. Noel, as the domestic mining industry declines, so does the enrollment in related educational programs. This is substantially reducing the number of qualified professionals in the mining industry. Is a comparable decline occurring for programs tailored to high-tech and manufacturing?

Mr. NOEL. This is kind of a subjective answer to your question, based upon observation, but I think if you go to many of our more outstanding technical universities, there is a predominance of people from outside the United States that go to these schools. And I think typically in the past what we have seen is that by and large they have relocated and they have come and they have actually become American citizens and have supported our industry.

The Forging Industry Educational and Research Foundation, we have 12 schools we work with. In discussions in Ohio State in particular, when you go to some of these schools, now as you talk to these students, their intent is no longer to stay here. Their intent is to go back to their native country and to work and develop industries of that country. And so I think everybody sees that the American youth is more—is interested in, really, pursuing jobs in the service segment of our country as opposed to get into the basic hard industrial sector. And many of us become very concerned about the ability to have the engineering capabilities to continue the advance of technology for both military and commercial aircraft.

Mrs. CUBIN. But the connection would be if jobs aren't available in the United States, why would students go into these type jobs and why would foreign students who are trained here stay?

Mr. NOEL. Well, I think the issue that you have here is there is no question there—you know, as far as there has, it has been a declining or consolidating industry, but one of the things that we—for example, you need a stable source of funding going into the technology sector. And with that, people are looking at the technology functions within their company, can stabilize their staffing, and they can go out and attract and retain people over long periods of time rather than be subjected to the economic cyclicality, which is you are bringing people in, you are laying people off, that sort of thing.

Mrs. CUBIN. Thank you. Now I recognize Mr. Gibbons for 5 minutes of questioning.

Mr. GIBBONS. Thank you, Madam Chairman.

Let me turn to Mr. Hanes first to ask a question, because I am curious on some of your testimony. When you hear some people say that we ought to be more reliant on recycling, the use of certain materials and metals and minerals that we produce, for example, hardened steel, that are included in communication processes or communication devices, how easy or how readily is our developed recycling process to take advantage of recovering those, to put those back in the market?

Mr. HANES. Well, if you start with the metal beryllium, of which I am most familiar, the manufacture of beryllium, it is based on recycling everything you possibly can. It is a very expensive commodity to win in the first place. But ultimately, you lose in the process the ability to recycle. Either technology no longer exists or it is not economically feasible or any number of reasons. If you put beryllium in copper to make beryllium copper, you can never get the beryllium back to make a beryllium metal component for a defense application, for example. If you put molybdenum in steel, you cannot get that back to put it into a superalloy, for example. You know, there are just basic laws of physics and chemistry that apply to that part of the process.

If you take the case of a precious metal inlay in a copper strip that is converted into a connector or a contactor that goes into an automotive ignition system in an automobile, the cost of recovering that precious metal, you know, back from inside of this very complicated structure is very high, although there are recycling companies that do that. But ultimately, you lose a percentage, you know, depending on the laws of physics, chemistry, and economics, you lose a percentage. So, you know, the concept, I think one of the favorite expressions of some of the opponents of mining is to say, well, if you need gold, you simply melt down all the jewelry. Well, eventually you run out of it and you have to find primary metal again.

So it is a lot different from industry to industry to industry, but the bottom line is it is all governed by, you know, physics, chemistry, and economics.

Mr. GIBBONS. And consumption. Some of these—

Mr. HANES. And consumption, correct. Thank you.

Mr. GIBBONS. What I would like to do is turn to Ann Carpenter and Mr. Silver, if I could, and propose this. When I was in Chile, I talked to Placer Dome's people there in their operation. And from the time they discovered their copper deposit until the time they had it into production was less than 2 years. What has been your experience with the time from discovery to the time of production of a mine in the United States?

Ms. CARPENTER. For example, I permitted a mine in 1997. It was the completion of it, roughly. I am not solid on that. But we did an EIS, an environmental impact statement, and it took us a year and a half. And it was a complex mine project, about a million and a half ounces, pit lake issues, a lot of environmental considerations and mitigation concerns. If I were to try to do that, just the permitting side, here, we are looking in the field at five to 10 years easy.

Mr. GIBBONS. Five to 10 years. And what is the average investment in the process before you ever get to production going through this permitting cycle and through the phase that you need to go through before you can stick the first shovel in the ground to get your first return on that investment? How much do you usually invest?

Ms. CARPENTER. Excluding litigation, the litigation variable—5, 10 million dollars.

Mr. GIBBONS. So you have $5-10 million just through the permitting process. That doesn't count for any of the development of, say, the mill site or the equipment—

Ms. CARPENTER. Or the mine, yes, right.

Mr. GIBBONS. Or the mine itself.

Ms. CARPENTER. Right.

Mr. GIBBONS. What if you included all of that by the time you got to the 10-year point?

Ms. CARPENTER. Well, for instance, I think the Glamus project, the Imperial project that was stalled recently even after it had gone through the permitting cycle, I believe they spent— if I might be able to ask a couple of my colleagues in the audience, if that is all right?

Mr. GIBBONS. Sure.

Ms. CARPENTER. Including capital costs, that estimate is $65 million.

Mr. GIBBONS. So they have got an investment of $65 million out there with an uncertainty, including a delay in the time to get that mine into production.

Ms. CARPENTER. Exactly.

Mr. GIBBONS. And as a result—

Ms. CARPENTER. Those are estimates that I would want to have the opportunity to look back on it.

Mr. GIBBONS. As a result, a normal business company is going to look at that and say, golly, this is something I have great questions about, whether or not I want to invest $65 in an uncertain future at an uncertain date, knowing that I could take the same $65 million and go down to South America, I could go to some other country and within 16 months have production and a return on that same money using the same environmental standards, using the same production and the same technologies that I would have used in the United States.

Ms. CARPENTER. Yes. I would say that.

Mr. GIBBONS. So it is discouraging. Mr. Silver, I want to ask you, too, the same question. And if you could, I know my time is up, but Madam Chairman has extended me the graciousness of going over my 5 minutes. The same type of a question, because your testimony very clearly alludes to the fact that it is the delays, the uncertainty, the legal cost adding up that make it unattractive for industry to remain in the United States. Now, I would like to hear.

Mr. SILVER. Yes, Chile is a good example, too, because in Chile mining has a primacy over all other land use. So it is a wonderful country for miners to go to.

What you see in most foreign countries is that there is a set path that is dictated by regulations. And as long as you keep within that path and you achieve the objectives and you submit the proper paperwork—

Mr. GIBBONS. Excuse me. So what you are saying is even in the United States, if you complete your permitting process to the standards that were required in the permitting and the environmental impact statement, there is no guarantee.

Mr. SILVER. There is no guarantee and it is actually, I think it is getting worse. You see a lot of people filing injunctions or trying to obfuscate the process at the 12th hour despite the millions that have been spent in the adequate time zones. And I think it is a very simple thing. You were mentioning before about budget cuts in the Federal Government. We are not asking the Government to spend a lot of money. We are saying just streamline the process. You can actually save money. Just make the process efficient and stick to it. Don't allow somebody to come in at the last minute and change the rules. Don't bend over backwards to give a special interest group another year, because you are destroying industry. And that is a risk that is unacceptable anywhere in business.

Mr. GIBBONS. One final thing that I want to ask about is the legal status. Oftentimes you see groups file lawsuits even though the environmental impact statement is clear in its purpose, clear in what you are going to do. The plan is set, it meets all the standards, and yet there is a lawsuit filed. What is the purpose of filing that lawsuit against you?

Mr. SILVER. Well, it is very simple. It is simple delay. If they can delay, it is expensive for the operating company. If you delay a project by 1 year, you will knock as much as 15 to 20 percent off its net present value. If they do that two or three times, the project is gone.

Mr. GIBBONS. So delay is —

Mr. SILVER. Delay is a tremendous tool for the opposition. But we have less rights than they do, because the delays are built into the system. We don't have the right to say the Government's decided this and they are sticking with it.

Mr. GIBBONS. I guess my final question—thank you, Madam Chairman, for your yielding me this—would be to Mr. Noel. What is your impression of the future of the mineral industry in the United States? Where do you think we are going?

Mr. NOEL. Well, to me the—you know, I am looking at what comes out of the mineral industry and how it goes into metals and what the future of the metals industry in the United States. And as far as the metals industry, you have key companies that have unique positions, like Timet and Henderson Nevada, that—you know, they have a $100 million sponge facility that is operating at 60 percent of capacity. And obviously one of the key issues there is absorbing their depreciation expenses on that facility. I mean, that is a huge concern. You know, at the current level of consumption, that is a business that is in trouble. You look at Special Metals, which is a primary producer of superalloys. They are in chapter 11.

So the question is, these are elements of both our commercial aviation and military defense systems. I believe we need this capability. And as this cycle prolongs, and the latest projections are that this probably will occur the last—you know, once the rebound and commercialization, they said 2005. And if they said ask me again next month and it is probably going to be 2006.

But I am really very concerned that these companies will survive in a fashion or form that will be able to support our needs. And I think it will be a huge loss to this country if that would happen. And they are—these are guys that are not—they are the prominent suppliers in their industries, and they are struggling at this point in time. So that is the reason we feel that the defense manufacturing supply chain has to work together. If we all work together and collaborate, you know, we can be more effective and more efficient.

And the key to survival is technology. And you look at what is the major issue, it is globalization. How do we respond to globalization? We feel it is technology, it is speed, it is collaboration, it is investment. So we see clearly a troubled sector. You go up the next step in the supply chain to the component producers, shape isn't as bad. They are generally at break-even points, but those businesses are going to be consolidated. Many of the smaller companies, I doubt they will survive the cycle.

So, I mean, it is a very troubling situation, with no clear path or no clear recovery point on the horizon.

Mr. GIBBONS. Well, I apologize to Mr. Cole for taking so much of his time that would be available to him. Thank you, Madam Chairman, for extending me the gracious time. Thank you.

Mrs. CUBIN. Certainly. I recognize Mr. Cole for 5 minutes.

Mr. COLE. Thank you very much, Madam Chairman.

Mr. Silver, your testimony was so nuanced that I want to probe it a little bit. I wish there were a lot more people here to hear what you had to say, quite frankly, because it was so much to the point. But let me ask you a couple of fairly simple questions.

First, I would assume from your testimony, obviously there is an enormous price differential producing abroad as opposed to producing here in a variety of areas in basic metals. How much of that differential, if you could categorize it broadly, is due to problems of permitting here as opposed to problems of supply? Or cost of production?

Mr. SILVER. That is a great question, because once the mines are built in the U.S., they are very competitive on a global scale. I mean, we do have higher labor costs, of course. Future environmental costs are built into the models, but we have better technology, we are more efficient, so we can keep our costs down and be very competitive at the global scale. So the producing side is not as big an issue as trying to be allowed to build the thing.

Mr. COLE. So you are pretty comfortable, if we could get the regulatory side under control, that we would both have the sufficient supplies and technology to be very competitive and essentially self-sufficient in a variety of areas?

Mr. SILVER. Yes. There is no evidence from all of our research that shows the U.S. is either picked over in the commodities we have traditionally done or that the future mines are going to be excessively expensive. The only evidence we see is this delay and the access.

Mr. COLE. Let me ask you this question and preface it with a comment. A lot of the—I know a lot about the oil and gas industry, not very much about hard metals extraction. A lot of the resistance in oil and gas that we meet is based on ideas about production that are about 50 or 100 years out of date. I mean, there really is sort of a vision in people's minds that we have a bunch of wooden oil wells up about 10 feet apart and that is what an oil field looks like today. And of course it doesn't. And we labor sometimes under the mistakes of the past. Well, we do have environmental problems—we have tremendous environmental problems around lead and zinc extraction in Oklahoma, a huge Superfund site there, where clearly what was done a century ago was an abomination, a waste of the resource.

Contrast, if you can for me, what the environmental consequences of hard metal are today, how much different and better the technology is compared to 50 or 100 years ago.

Mr. SILVER. OK. One of the things I find most interesting is that the people who don't like the mining industry assume we are using 100-year-old technologies. And in fact, if you use old technologies, it is expensive, because you don't fix the problems. The new technologies that are being developed today are light years ahead of what it was 100 years ago. And as a consequence, the industry spends a lot of effort trying to use better technology because it in fact reduces cost. So the argument that we don't use modern technology is just simply not true, and it is not to our advantage to not use the best available technology.

Mr. COLE. Let me direct this question, if I may, to Ms. Carpenter, although any of you would be—I would be delighted to hear from you.

Clearly, I mean, there are sort of competing environmental visions that you run into. One is everything needs to be preserved pristine. The other is let's use what we need to use, but let's do it responsibly and let's restore it once we are done.

Tell me just broadly where the industry is at in terms of its ability, once an area is tapped out, so to speak, in terms of productive capacity, to restore or otherwise minimize whatever environmental consequences where they are actually mining a particular metal.

Ms. CARPENTER. Very good question. And again, it lends back to the technological advances. I work quite a bit doing reclamation work on projects for clients that are leaving this country and going elsewhere. So they are closing their projects out. And I share a lot of my data across the sectors in the mineral industry, including construction materials, the coal industry, hard rock. And the advances that people are doing, the advances, the technological advances, the scientific advances have all led to incredible reclamation efforts and much better than we saw even 10 years ago. And I would offer that, with the advances in technology and the understandings that we see and the idea-sharing, we are coming up with much, much better reclamation efforts.

And it is a constant stream of—again, another dynamic area—where we are always seeking for better efficiencies on it. It is toward the bottom line, but we also want a better final product.

Mr. COLE. Let me ask you this question. Would it be fair to say that, you know, once we get past the permitting processes and into production that this country is actually a leader in terms of restoration and—you know, after an area is exhausted?

Ms. CARPENTER. It is absolutely true. Wherever I have worked overseas, they have commented exactly on that.

Mr. COLE. Well, could you argue that, but would it be fair to argue? It might be pushing the point, but would it be—and don't hesitate to disagree. It would be fair to argue, you could make the case it is environmentally more responsible to mine here because—you know, if our policies drive us to other countries that don't have the technology or the commitment to deal with the environmental problems, then you are going to have, frankly, in the global sense, worse environmental damage in some of the places that you are working than you would have if you worked inside this country where we have the resources, the technology, and frankly now, the ethos to make sure the environment stays relatively, you know, undamaged.

Ms. CARPENTER. I would encourage you to get on the Sacramento Bee website and look for a Tom Knudson article called "State of Denial." And it addresses that in particular. In his article, he shows, or sort of illustrates that preservation stops at the border. And by exporting our production, our consumption needs elsewhere, we can't do it to the same quality that we can here.

I would suggest, and I believe this strongly, that we should be exporting our knowledge and proving—we are environmental—we are leaders in the world on our technologies, in this industry and others, and in our environmental successes and regulations. And they find it pretty hypocritical that we keep exporting out our—wherever I have worked, anyway—our consumptive needs when they know that we can do it to such a high level at home. And they strive for those high levels.

Mr. COLE. I see my time is probably up, Madam Chairman. Thank you very much.

Mrs. CUBIN. Thank you. I would like to make a statement that before Federal policy can ever be changed or moved, we have to identify the problem and we have to show people in very concise terms that there is a problem. And one way we could do that is for you to help us identify a mineral or some minerals that are critical to our national defense or to commercial aviation or something, and really as many as possible, that in the short term are really going to cause a problem for us. And if we can demonstrate this to the administration, then possibly we can get a national mineral policy on their radar screen and we can begin to elevate the importance of this issue. Because I am convinced it is very important. And I would be glad to work with Mr. Gibbons, Mr. Cole, anybody on the Committee to pursue this further.

So I would appreciate it if you could help us with that. Mr. Gibbons?

Mr. GIBBONS. Madam Chairman, I had just one statement. Of all this testimony that we have received before this Committee, this is probably some of the most important with regard to our mineral industry and where we are going.

I think one of the things we have forgotten about the Governmental agencies that deal with these industries, and many times we have seen and I am sure some of these companies have seen some of our own internal Federal and State agencies derail permitting processes through coordination with organizations who are opposed to mining and other resource development. And we need to bring in those agencies and find out the underlying premise and the issues that they are dealing with when it comes down to their opposition to the mining industry as well.

We are seeing the industry here come up and talk about their frustrations, and I think we need to deal also effectively with our own Governmental agencies that add to that frustration.

Mrs. CUBIN. That, and additionally the reticence of an agency to make a statement about policy when in fact they are very aware that certain policies might be beneficial.

I would like to make an announcement. John Rishel, who worked for this Committee and was passionate about national mineral policy and he was passionate about making mining for hard rock minerals in the United States more available and more profitable, while at the same time protecting the environment—there is a memorial for John today at 4 p.m. in Longworth Room 1334, this room. So anyone who would like to attend that would certainly be welcome. And I think John certainly is deserving of a memorial to bring out the contributions that he made. So thank you very much.

I thank the witnesses for their valuable testimony and the members for the questions. Once again, members of the Subcommittee may have additional questions, and we would ask that—we will get them to you and ask that you could respond within 10 days, at which time the record will close.

If there is no further business in front of the Subcommittee, then we stand adjourned.

[Whereupon, at 11:48 a.m., the Subcommittee was adjourned.]

[Additional information submitted for the record by the Mineral Policy Center follows:]

Statement of Lexi Shultz, Legislative Director, Mineral Policy Center

The Antiquated 1872 Mining Law Subsidizes Environmental Destruction and Public Health Risks:

There are several ways in which the outdated 1872 Mining Law has, by subsidizing the mining industry, actually encouraged environmental destruction. First of all, by selling off public lands at below market prices and letting the industry have free access to $240 billion worth of public minerals, the law has blocked taxpayers from getting a fair return on their resources and encouraged the mining industry to overuse public lands.

Second, by failing to provide land managers with the discretion to deny mines once valid mining claims have been established, the law has elevated the interests of the mining industry over all other possible uses and interests in the land, including those of taxpayers, other businesses benefiting from the land, public health and the environment. That is, under the 1872 Mining Law, federal land managers must approve mining operations with valid claims, even if tourism would be a more economical use of the land, even if drinking water and public health would be compromised by the presence of a mine in the area, and even if the land is of particularly crucial significance for the health of an ecosystem.

Third, by omitting any standards to address the specific environmental problems caused by mining, the law has failed to prevent mining-related pollution, leaving a legacy of 12,000 miles of polluted streams and rivers, air pollution, contaminated drinking water, and disrupted habitat. While the mining industry must comply with the same environmental laws that every other business in America must comply with—the Clean Water Act, the Endangered Species Act and others—these laws do not protect against contamination of groundwater from heavy metals and chemicals like cyanide. Nor do these laws directly address acid mine drainage caused by rain and snow falling on mining wastes, or the metal-laden dust and particulate matter created by mining operations. Considering that modern mining produces more waste than all other sources combined, including municipal waste, and that mining contamination can last for hundreds or thousands of years, it is inexcusable that everything possible is not being done to prevent this pollution from occurring in the first place.

Fourth, the law lacks any standards either for the reclamation of mined-out lands or for the mining industry to bear the financial responsibility from that cleanup. As a result, the law has encouraged mining companies to simply walk away from mines once they are done, leaving taxpayers with devastated public lands and an enormous cleanup bill. The Mineral Policy Center has estimated that taxpayers will eventually have to pay $32 to $72 billion to clean up the more than half a million abandoned and polluting mine sites across the country. Seventy of these sites have been designated as Superfund sites because of the enormous damage to human health and the environment that they are causing. Each of these sites alone may cost in excess of $100 million to clean up. Moreover, because the mining industry knows that it will not be liable for the full costs of cleaning up abandoned mines, it has no incentive to minimize the amount of pollution it will produce in advance.

The Consequences of Mining on the Environment and Human Health have been Devastating:

The consequences of the 1872 Mining Law's outdated and misguided policies have been devastating, both to the environment and to human health around the country. Consider the following examples taken from a report produced by the Mineral Policy Center entitled "Burden of Gilt."

In Montana, windblown particulates from old mine tailings piles in and around Butte deposited heavy metals on high-school baseball fields in such dangerous concentrations that the fields had to be excavated and the topsoil had to be replaced. The city's water treatment plant is built on old tailings deposits that contain dangerously high concentrations of copper, zinc, cadmium, arsenic, and lead. Sediments from mine tailings have contaminated more than 35 square miles of groundwater in the Butte area.

In Idaho, lead levels in Silver Valley soil downwind from the abandoned Bunker Hill silver mine—designated as a Superfund site—were found to be more than 30 times higher than maximum levels deemed "safe" by the Environmental Protection Agency. Virtually all of the 179 children living within a mile of the site were found to have potentially brain-impairing levels of lead in their blood.

In New Mexico, after a molybdenum mine near Questa was inactivated when prices fell in the mid-1980's, tailings from the mine continued to contaminate the Red River, killing fish and destroying wildlife habitat, and also contaminated wells relied on by Taos County residents.

In Colorado, the Summitville cyanide-leach gold mine sent a flood of cyanide and heavy metals into the Alamosa River, killing fish and destroying 17 miles of the river, when its 45-acre waste pile flooded and leaked. The Galactic Resources Mining Company paid only minimal fines and a $4.7 million performance bond before filing for bankruptcy. The state and the EPA have already spent $130 million to clean up this Superfund site, and estimate that the job will take another $45 million of taxpayer money to complete.

In Montana, the Pegasus Gold Corporation went bankrupt, leaving taxpayers to pick up the approximately $100 million bill to clean up the now defunct Zortman-Landusky mine.

The Antiquated 1872 Mining Law Must be Comprehensively Reformed:

It is clear that, under the current provisions of the 1872 Mining Law, far too much environmental destruction and far too many public health risks have occurred. In order to prevent these horror stories from being repeated, and in order to protect public lands and the Americans who own those lands, the entire 1872 Mining Law must be reformed. This reform must be undertaken comprehensively and with the goal of protecting the interests of taxpayers and the environment, not just those of the mining industry. Accordingly, the law must be reformed with three basic principles in mind:

Fair Return to the Taxpayers. The mining industry should be required to pay fair market value for both public minerals and public lands. For minerals, the mining industry should pay a 12.5 percent royalty, which is the same as the royalty paid by the oil industry for drilling on public lands and by the coal industry for mining above ground on public lands. Since mining companies pay royalties for mining on both state and private lands, there is no reason why they cannot give federal taxpayers the same consideration. For public land, the practice of "patenting," selling land for $2.50 to $5.00 an acre, should be permanently abolished. Mining companies should be forced to engage in arms-length transactions with the federal taxpayers, just as they would if they were dealing with private entities. The taxpayer-subsidized "free lunch" must end.

Pollution Prevention. The best way to protect clean water, clean air, and wildlife habitat is to prevent pollution in the first place. Reform of the law must include environmental standards that will address the specific environmental problems caused by hardrock mining, including groundwater contamination, runoff, and other such problems. These standards must also deal with the reclamation of mined-out lands so that these sites do not become a hazard in the future. Moreover, where no amount of environmental safeguards could adequately protect a particular region, whether because of its importance for drinking water, for wildlife habitat, for recreational opportunities, or any other reason, land managers must be given the discretion to deny mines in order to protect those interests. Finally, in order to minimize the potential for destruction, the mining industry should not have access to unlimited amounts of public land for the dumping of mining wastes.

Polluter Pays. The mining industry must be required to pay for the clean-up of both depleted mines and any mines for which pollution prevention failed. These costs should be borne by the industry that caused the pollution or degradation of the land. In this way, taxpayers will not have to shoulder what should be a normal cost of doing business for an industry taking advantage of public resources, and the mining industry will have an incentive to mine as responsibly and with the least disruption as possible.

The Mill Site-Mine Site Issue is the Law and Must be Enforced:

There is one provision in the 1872 Mining Law that has the potential to protect taxpayers and the environment, by limiting the amount of public land available for the dumping of toxic mining wastes. This provision provides that mining companies may have a mill site of 5 acres, to be used to process or dump overburden and ore, for each 20-acre mineral claim. Despite industry claims to the contrary, this provision has been a part of the law since 1872. And despite industry claims to the contrary, the law does not allow unlimited numbers of mill sites for each mineral claim. If it did, there would be no need for the law to have limited the size of mill sites to 5 acres—it could simply have provided that mining companies be allowed to have a mill site of unlimited size.

The fact that this portion of the law has been inconsistently enforced until recently is highly unfortunate and may have led to otherwise avoidable environmental degradation. However, this fact does not alter the provision's legal authority.

Nor does this provision mean the end of mining, as the industry claims. Mining companies can do land exchanges, apply for special use permits, or, in some instances, reconfigure their mines to mine underground. While these alternatives may not be as favorable to the mining industry as having access to unlimited amounts of public land at little or no cost, they are the only options that can even begin to give taxpayers and the environment a fair shake.

Because the law is finally being enforced the way it was written, the industry may try to push weak or "sham" reform. Mineral Policy Center opposes any attempts at reform of the 1872 Mining Law that do not adequately address the needs of taxpayers, the environment, and public health.

Conclusion:

It is time to put an end to the taxpayer-subsidized environmental destruction and public health risks caused by the outdated 1872 Mining Law. What is needed is meaningful comprehensive reform of the law that will fully protect the interests of taxpayers and the environment.

Statement of Stephen D'Esposito, President, Mineral Policy Center

Mineral Policy Center Supports Responsible Mining Polices and Practices

Mineral Policy Center supports responsible mining policies and practices: responsible mining policies that give taxpayers a fair return for valuable land and mineral assets, and that eliminate government subsidies to mine on public lands; responsible mining policies that require mining companies meet adequate environmental protection standards; and responsible mining policies that recognize that on some public lands there are resources, and other uses, that may be more valuable than mining, including the protection of environmentally significant areas. Currently, federal law does not offer adequate environmental or taxpayer protections.

Mineral Policy Center recognizes that some mining companies seek to operate in a manner that protects our environment. But the 1872 Mining Law is actually a disincentive for responsible industry action. The 1872 Mining Law sends the wrong signal to mining companies. It rewards irresponsible behavior. Until it is reformed it will serve as a haven for bad actors and fail to reward those who act responsibly.

Those who most vociferously oppose environmental reform of the 1872 Mining Law, may be those companies who do not believe they are poised to operate successfully in an environment that mandates and rewards environmental performance.

Successful mining law reform will balance the interest of mining companies with those of taxpayers, citizens who seek to protect land and water resources, and future generations who will benefit from well managed public lands.

131 YEARS LATER, ITS TIME FOR MINING LAW REFORM

One hundred and thirty-one years is too long. It is time to reform the 1872 Mining Law. Written to encourage the development of the mining industry, and the settlement of the western United States, the mining law is a relic of a bygone era—a time when mining was a pick-and-shovel affair, when men moved on horses and in covered wagons.[1] The mining law's roots may also have a humanitarian origin, to deter violence that resulted from claim jumping.[2] One hundred and thirty-one years after its passage, its original purposes accomplished, it is time for reform.

Today, the prospector's pan has given way to giant earth-moving machines that cam literally crumble mountains and dig pits the size of small towns. Panning for gold nuggets has given way to the use of potentially lethal chemicals such as cyanide that leach microscopic flecks of ore from massive piles of pulverized rock. Today's prospectors are multi-national corporations and their mine sites occupy thousands of acres of our public land.

While there are technical and engineering solutions to many of the environmental problems that mining can cause, technical solutions are only part of the answer. They will not be enough to fully address the broader environmental, economic, social, and cultural issues that this Subcommittee, and all Members of Congress, must grapple with. Reforming the mining law is not, after all, just a matter of a technical fix. It is one thing to design a safe and efficient mine, it is quite another to design public policy that results in good decisions about the use of public land and resources. Good public policy must provide a basis for weighing environmental, social, economic, and cultural issues, as well as technical issues.

MILLSITE CLAIMS AND THE 1872 MINING LAW

With the enforcement of the millsite limit of the 1872 Mining Law, it appears that we may have a new ally in the fight for mining law reform, the mining industry.

Organizations Representing Millions of Members Support the Millsite Decision

MPC believes the millsite opinion, effectively enforcing limits on mine waste dumping on public land, is based upon an accurate reading of the 1872 Mining Law. While it is widely accepted that the Mining Law's millsite restriction does not meet the needs of some of today's mining operations, neither does this antiquated law meet the needs of taxpayers, communities near many of today's mines, or the environmental health of our public lands. The massive waste piles produced at many of today's mines have outgrown the mining law.

However, the solution to this problem is not to create a special exemption from the part of the mining law that some in the industry find troublesome. The solution is comprehensive reform that will balance the needs of the industry, taxpayers and the environment. We should engage in a public debate about reforming all of the mining law, not just the part that the mining industry does not like.

In a March 23, 1999, letter to Secretary of the Interior Bruce Babbitt and Forest Service Chief Michael Dombeck, Mineral Policy Center, Friends of the Earth, Natural Resources Defense Council, Sierra Club, The Wilderness Society, U.S. Public Interest Research Group, Okanogan Highlands Alliance, Western Organization of Resource Councils, and Grassroots Environmental Effectiveness Network (GREEN) petitioned the government to reject the Plan of Operations for the Crown Jewel Mine because it was over the millsite claims limit. To quote from the letter: In this case, the federal land agencies must determine whether to approve a mining plan that is proposed on public lands that do not contain valid mining and millsite claims under the 1872 Mining Law. A number of other proposed open pit gold mines on federal land face similar issues. The most pressing examples include the Imperial Project in southern California and the Yarnell Mine adjacent to the town of Yarnell, Arizona. Thus, your decisions at Crown Jewel will have ramifications across the West.

On March 25, 1999, the U.S. Departments of the Interior (DOI) and Agriculture (DOA) released a joint decision stating that they were "unable to approve the proposed Plan of Operations" for the Crown Jewel Mine in the State of Washington.[3]

The plan for this large open-pit, cyanide-leach gold mine was rejected because it did not comply with the requirement of the 1872 Mining Law that limits claimants to one 5-acre millsite claim for each mining claim.

The March 25, 1999 decision references the November 7, 1997, Solicitor's Opinion entitled "Limitations on Patenting Millsites Under the Mining Law of 1872." This 1997 opinion reviews the millsite limit in detail. The conclusion is unequivocal: "the plain language of the mining law indicates that only one 5-acre millsite claim per mining claim may be patented." [4]

Before the March 25th decision, mining companies were sometimes permitted, albeit illegally, more than one 5-acre millsite claim per mining claim. Although the Bureau of Land Management (BLM) and U.S. Forest Service (USFS) have permitted multiple millsite claims in some instances, they had no legal basis under the mining law, or under regulations, for such approvals.

The Solicitor's Opinion notes that the current BLM's Handbook for Mineral Examiners may be the source of the problem. It provides for the granting of multiple millsite claims per mining claim. However, as the Solicitor makes clear, "no authority is cited for these statements." While this explains why some BLM field staff may have approved plans of operations that were over the millsite limit, it does not change the fact that the limit exists in the Mining Law.

This is Not Just About Land Use

The recent millsite ruling addresses a fundamental environmental problem—today's mines are dramatically larger and produce more waste than the pick-and-shovel operations that the mining law was written to govern. Management of this waste presents a significant public policy challenge since waste from mines often pollutes surface and groundwater resources with acid mine drainage and heavy metals such as arsenic. The Mining Law does not address these environmental concerns directly. To the extent that it deals with them at all, it addresses them through the millsite limit. This de facto environmental safeguard should not simply be jettisoned, it should remain in place until the Mining Law is reformed to include such protections by design.

History Supports the Millsite Decision

There is ample evidence that Congress, DOI, and many in the mining industry understood the strict millsite limit contained in the Mining Law. Prior to 1960, the Mining Law allowed millsite claims only in connection with vein or lode claims, not with placer claims. In 1960, Congress explicitly amended the mining law to permit the location of millsites in connection with placer claims. [5] The legislative history of the amendment makes it clear that Congress and DOI understood both the millsite limits in the existing statute and the amendment, which permitted only one five-acre millsite claim in connection with a placer claim. The amendment was passed and signed only after input from the DOI caused Congress to remove language that would have permitted millsites "for each individual claimant" and allowed for a larger millsite claim of "10 acres for each individual claimant." [6] The amendment was changed as requested by DOI, which sought to have the millsite limit for placer claims match that of lode claims. The report language was explicit:

> [T]he word "ten" was stricken and the word "five" inserted in lieu thereof. *The purpose of this amendment is to restrict the area of a millsite in conjunction with a placer claim to 5 acres of land to make it conform with the allowable millsite acreage for lode claims which has been the statutory requirement since 1872 ...*
>
> *[T]he words "for each individual claimant" were stricken so as to impose a limit of one 5-acre millsite limit in any individual case preventing the location of a series of 5-acre millsites in cases where a single claim is jointly owned by several persons ....*

In 1968, the leading mining industry trade association (the American Mining Congress, the predecessor of today's National Mining Association), in a statement submitted to the Public Land Law Review Commission, acknowledged that the law permits only one millsite claim per mining claim. "When the mining law was enacted in 1872, provision was made for the acquisition of five-acre millsites to be used for plant facilities on mining claims. The typical mine then was a high-grade lode or vein deposit from which ores were removed by underground mining. The surface plant was usually relatively small, and acquisition of five-acre millsites in addition to the surface mining claims—adequately served the needs of mines "Today, the situation is frequently different—A mine having 500 acres of mining claims may, for example, require 5,000 acres for surface plant facilities and waste disposal areas. It is obvious that such activities may not be acquired through five-acre millsites." [7] (emphasis added)

In 1974 in United States v. Swanson, 14 IBLA 158 the Interior Board of Land Appeals stated that: [A] claimant is entitled to receive only that amount of land needed for his mining and milling operations, and this amount can embrace a tract of less than five acres. The statute states that the location shall not "exceed five acres." ... The reference to five acres in the statute is clearly a ceiling measure, not an absolute, automatic grant.

In 1979, the Congressional Office of Technology Assessment confirmed this interpretation of the mining law's millsite provision. There could be as many millsites as there are mining claims, and each millsite would be at most one-fourth the size of the typical 20-acre claim, so that millsites, in the aggregate, would be one-fourth the size of the ore body encompassed by the claims.

The Impacts of Enforcing the Millsite Opinion

Will this be the end of mining on public lands? No, there are mining methods, such as underground mining, that do not necessarily require such vast amounts of space for processing and waste dumping. It is also important to remember that open-pit mining takes place on non-federal lands. In those instances, mining companies successfully assemble the necessary lands through acquisition, leases, or by purchasing mineral rights. As the Solicitor's Opinion made clear, mining companies can seek to acquire necessary millsite acreage through land exchanges and special use permits. Both methods do require the company to meet additional hurdles and land managers could exercise discretion to prevent "unnecessary" or "undue" degradation of public lands. But these are not unreasonable conditions.

The decision may have an impact on marginal, low-grade deposits. But there is no reason why federal policy should be used to subsidize the mining of such deposits on public lands. In fact, it should be the government's policy to create a level playing field for all mining companies—whether the land is owned by the federal government, the state, or private citizens.

We do not expect open-pit mining to end on public lands as a result of this opinion, nor would we expect, does DOI or the Forest Service.

The Millsite Limit Should be Enforced Until the Mining Law is Reformed

We do not think that the law should be effectively amended, and weakened, to suit the needs of a number of mining companies or even an entire industry. We should not look at this problem from the wrong side. The underlying problem is not the millsite limit. The millsite limit is a symptom. The problem is the outdated 1872 Mining Law. It does not fit today's mining industry, it does not protect taxpayers, and it does not protect the environment. Let's fix the underlying problem, not just treat the symptom.

Some believe that our reading of the millsite limit under the 1872 Mining Law is legally flawed. We disagree. But for those who hold this view there is an obvious remedy and it is not this Subcommittee or the Congress. It is the courts.

Some believe there is an urgent need to address this issue because of concerns about retroactivity. We do not think a rush to judgement on these important issues should be driven by such concerns, and we do not believe this issue should be applied retroactively. If necessary, we would support a resolution or amendment specifying that there would be no retroactive application of this issue to currently operating mines. Our objective is not to penalize mining companies that may have benefited from the incorrect application of the Mining Law by either DOI or the Forest Service. Our objective is that this provision now be applied.

WHAT'S WRONG WITH THE 1872 MINING LAW

The 1872 Mining Law allows for public land giveaways and taxpayer subsidies to the mining industry. And it fails to protect our environment and our public lands. Of course, those who crafted the mining law in 1872 could not envision the potential environmental impacts of modern mining. The environmental legacy of this outdated law is all too clear. A 1989 report from the U.S. General Accounting Office found that the Mining Law "runs counter to other national natural resource policies and legislation." [8] The GAO found that the mining law "no longer promotes mineral development" and can result in "needless damage" to public lands. [9]

Let's Cleanup Our Nation's Abandoned Mines

Estimates of abandoned mines, range from at least 200,000 to over 500,000, scattered across the country. The abandoned mine problem should serve as a wake-up call. Cleanup costs could be as high as $72 billion. [10]

We should immediately implement a national program to clean up the hundreds of thousands of unreclaimed and abandoned mine sites. MPC estimates that there are 557,000 abandoned mine sites nationwide, with an estimated cleanup cost of $32 to $72 billion. The Western Governors Association has identified abandoned mine sites in 10 states in need of priority cleanup. In Alaska, the abandoned Treadwall Mine is pockmarked with vertical shafts, open portals and pits, and a dangerous highwall 500 feet tall. Because the site is adjacent to the cities of Juneau and Douglas it receives extensive use by the public. Washington state residents are struggling with a toxic legacy at the Holden Mine where acid mine drainage from waste rock and 18 million tons of tailings have rendered 12 miles of nearby river biologically dead. In Utah, the abandoned Temple Mountain site is home to 300 open uranium mines with moderate to high radionuclides.[11] Californians are still living with the festering Iron Mountain Mine, which is predicted to continue leaching acid for at least 3,000 years. And that is just to name a few.

Reclamation of sites like these would restore valuable lands, eliminate public safety and health threats, and create up to 10,000 jobs.[12] The question is not whether it should be done, but how it should be funded. Potential sources of funding include a mineral royalty, rental fees, and through other sources such as a reclamation fee. Establishment of this fund should not be delayed.

We Should Permanently End $2.50 An-Acre Public Lands Giveaways

Although an annual moratorium is in place, the 1872 Mining Law allows public land giveaways at the bargain-basement rate of $2.50 or $5.00. This was a bargain in 1872, today it's a steal. Patenting is not necessary to mine on public lands.

According to the Department of Interior, during the Mining Law's first 120 years, 315 million ounces of gold, 5.5 billion ounces of silver, 79.5 million tons of copper, 19.2 million tons of lead, and 13.9 million tons of zinc were mined in 13 western states. In 1994 we estimated the value of these minerals at more than $231 billion. That's just the minerals under the land, it doesn't take into account the value of the land.

Taxpayers Deserve a Fair Royalty

When a mining company mines on public land, they do not pay a royalty. When they mine on private or state land, mining companies pay a royalty that can range from 5% to 18%. What is the justification for not requiring a royalty for mining on public lands? There is none. The coal, oil and gas industries pay a fee when they mine on federal lands. Why not the hardrock mining industry?

Consider these excerpts from the testimony of Dr. W. Thomas Goerold, a noted minerals economist to the Senate Energy and Natural Resources Committee, Subcommittee on Mineral Resource Development and Production, on September 13, 1990:

"Current domestic hardrock mineral producers sometimes claim that paying for federal minerals would be so burdensome that it would force a significant portion of them out of business. A cursory examination of the evidence does not support these claims. Producers of leasable minerals found on federal lands have paid royalties and land rentals since 1920 and no one questions the health of these industries. Moreover, miners of hardrock minerals have a long history of routinely paying royalties and rental payments when these same minerals are found on state or private lands."

"Hardrock mineral miners maintain that there is still a fundamental difference between hardrock minerals production and other businesses, as well as between hardrock minerals firms and other mineral producers that pay land rental and royalty fees to the Federal Government for use of publicly owned resources. Contrary to industry claims, these purported distinctions do not justify the privileged treatment accorded producers of hardrock minerals. The Office of Technology Assessment supports this view. The OTA believes that the distinctions between leasable (generally energy and chemical minerals requiring government permission and payment of lease and royalty fees) and locatable minerals are more artificial than real."

Do hardrock miners on federal lands have more importance than automobile manufacturers, retail store owners, or any other business not eligible for similar government subsidies? Are hardrock miner producing minerals from federal lands more important than these same producers mining state or private lands?

One argument advanced by mining interests against the imposition of royalties for federal hardrock minerals is that the Federal Government already taxes the profits of these companies. This is a misleading argument—most non-mineral businesses do not obtain the inputs to their firms from the federal government at no cost, yet virtually all pay a federal income tax. Royalty and rental free mineral operations are analogous to a gift of steel and rubber to automobile manufacturers, or free office rental to an accounting firm, courtesy of the U.S. Government.[13]

There are also federal land parcels in Minnesota, Missouri and Ilinois where miners pay royalties for extraction of hardrock minerals. And even on federal lands, mining companies are willing to pay royalties, to other mining companies but not to the taxpayer. [14] In October 1993, Newmont Mining Corporation leased 1872 Mining Law claims on BLM Land at Grassy Mountain in Oregon from the Atlas Corporation. Newmont paid a $22.5 million cash bonus and a $5 net smelter royalty production.

A net smelter royalty of between 8% and 12.5% should be enacted. The proceeds from this royalty should be used to fund abandoned mine cleanup.

We Should Let Mining Compete With Other Land Uses

131 years ago it may have been possible to make a case that mining deserved preferential treatment on public land, over all other uses. Today there is no social or economic good that justifies this preferential treatment. There are public lands that deserve protection, and there are public lands that are more suitable for other economic or recreational purposes.

Land managers should have the authority and discretion to protect environmentally significant public lands, weigh other land uses, and deny permits for poorly planned mines. A mining permit application must clearly demonstrate, before mining begins, how the mining and reclamation project will occur so as to minimize environmental impacts. Land managers must have the authority to deny mining permits in environmentally fragile areas, or critical wildlife habitats and areas otherwise found to be unsuitable for mining.

We Should Protect Our Public Lands and Environment

Environmental safeguards must protect water resources and prevent significant long-term environmental damage. It is worth noting that a 1987 study by the EPA rated problems related to mining waste as second only to global warming and stratospheric ozone depletion in terms of ecological risk. The report concludes "with high certainty" that the release to the environment of mining waste "can result in profound, generally irreversible destruction of ecosystems." In a 1985 report the EPA stated that mining for hardrock minerals, asbestos, and phosphate alone generates 1 to 2 billion metric tons of waste each year, and that "perhaps 56% of the waste generated could be considered potentially hazardous to human health or the environment." Mining has polluted 12,000 miles of rivers and streams and 180,000 acres of lakes. [15]

Environmental safeguards should include provisions for water protection, full cleanup and reclamation, environmental operating standards, enforcement requirements, and guarantees that the mining company will pay for full closure and cleanup. We are all familiar with the Summitville nightmare. In 1992, the Summitville Consolidated Mining Company declared bankruptcy and walked away from its environmentally disastrous cyanide, heap-leach gold mine in the San Juan Mountains of southern Colorado. Taxpayers, meanwhile, have been left with a devastated landscape and an enormous cleanup bill. So far the State and EPA have spent $130 million dollars to reclaim and restore the site, and expect to spend another $45 million dollars before the job is done. [16] With a reclamation bond of $4.7 million [17], taxpayer liability equates to approximately $170.3 million. This is today's problem because the taxpayer bill is still mounting.

The Public Should Participate in Mining Decision on Public Lands

The public must have the right to fully participate in mine decisions on public lands. This includes access to information, the right to comment on permit and regulatory actions, the right to petition the government to designate an area unsuitable for mining, the right to file citizen complaints, the right to accompany an inspector to a site, and citizen suit provisions to compel enforcement. The public must have the right to fully participate in mine decisions on public lands.

ZORTMAN-LANDUSKY, A REFORM CASE STUDY

The Zortman-Landusky Mine (ZL) is a prime example of why the 1872 Mining Law needs urgent reform The Zortman-Landusky mines is located on BLM managed lands in the Little Rocky Mountains of Montana. It is the home of the Assiniboine and Gros Ventre Tribes of Fort Belknap who have a strong cultural and spiritual connection with the Little Rockies. In fact, Spirit Mountain, where the mine site is located was considered a sacred site.

This was a large-scale open-pit cyanide heap leach gold mine. It was the largest gold mine in Montana and it mined the lowest grade ore in the United States. It caused more land disturbance per amount of gold extracted than any other gold mine in the U.S. During typical operations, more than 55 tons of ore are processed to yield a single ounce of gold.

During its operation, Zortman-Landusky experienced a litany of cyanide solution leaks and spills, stability failures, surface and groundwater contamination, bird and wildlife fatalities, and other problems. Streams emanating from the mine area, including water flowing onto the reservation, were seriously polluted with acid and heavy metals. The mine experienced numerous cyanide spills. There have been severe problems with acid streams and fish kills. Following a major spill, cyanide appeared in domestic drinking water in a mine worker's housing unit south of the mine.

The mine was operated by Pegasus Gold Incorporated. Responsibility for the mine's troubled track record rests not only with Pegasus Gold, but the 1872 Mining Law and weak environmental safeguards.

Things got worse when the stock of Pegasus plunged in late 1997 due to the falling price of gold and legal troubles. Pegasus declared chapter 11 bankruptcy on January 16, 1998.

Today in the state of Montana, government officials estimate that this recently abandoned open-pit cyanide heap-leach gold mine could cost state taxpayers $8 million in cleanup costs.[18] An independent mining engineer has estimated that the cleanup bill to taxpayers could be an additional $90 million. It is estimated that Pegasus produced $360 million worth of gold at the mine and returned $0 to the taxpayer. And it is likely that the taxpayer will get stuck with a substantial cleanup bill.

During the bankruptcy proceedings, the company outraged both state officials and the public when it sought to pay executives a bonus of $5 million, just as state officials revealed that taxpayers could get hit with a cleanup bill of $8 million. And in June the company appointed by the state to handle reclamation, was fired because it had already spent its entire annual budget in just six months.

There has been a history of problems like this with cyanide process gold mines in Montana. As a result, in 1998, Montana voters passed a ban on all new cyanide process mines.

NOW IS THE TIME FOR COMPREHENSIVE MINING LAW REFORM

It has been ten years since the last significant attempt was made to reform the mining law. There is too much at stake on our public lands, in our Western communities, for American taxpayers, and for the mining industry, to delay further.

Over the years, a number of bills have been introduced under the reform label that are actually not true reform bills because they fail to adequately address the fundamental environmental and fiscal shortcomings of the mining law. These bills typically contain miniscule royalties and wide loopholes for escaping royalty payment, fail to address important environmental protection issues, and do not allow land managers to weigh other uses of public lands. One telltale sign of a sham reform bill is the use of a "net proceeds" royalty. The "net proceeds" royalty allows companies to deduct so many costs before paying a royalty that the taxpayer ends up with almost nothing.

Some will argue that now is not the time to reform the mining law because mining companies are already suffering as a result of today's low mineral prices. But mineral prices have, and always will, fluctuate. It is in the public's interest to take action that will stimulate other commercial and non-commercial uses of our public lands, including preservation. And it is in our best interest to pursue environmental objectives that will lead to job creation in mining communities or former mining communities, such as abandoned mine cleanup. Taxpayers deserve a fair royalty and our public lands deserve environmental safeguards, whatever the price of metals happens to be.

Policies that provide public subsidies to mining companies create an incentive for inefficient mine operations on public lands, perhaps in places that are best used for other purposes. These subsidies lead to an unfair economic advantage for some companies and may result in inefficiencies and over-supply. The net impact of such policies is to make mining more attractive on federal lands than on other lands. "The Federal government, by forgiving this normal mineral business cost, has distorted the distribution of economic activity, discouraging mining on private, state, and tribal land and encouraging it on Federal land."[19] Continuation of this policy is not in the country's economic interest.

It is time to put an end to the subsidies and favors that mining companies receive on public lands. The net results of 1872 Mining Law reform will be healthier communities and healthier ecosystems, jobs creation, and, we believe, a healthier mining industry.[20]

A mining industry that is rewarded for its environmental performance, and penalized for its environmental mistakes, will be a healthier industry, both in the United

States and around the world. It is in the interest of this Subcommittee to create incentives for better environmental performance on our public lands. Improved environmental performance will increase the competitiveness, marketability, and performance of U.S. mining companies.

The United States economy has developed to the point that mineral development no longer needs preferential treatment on our public lands. The way we manage and use our public lands today will have an impact on the landscape and opportunities we pass on to future generations. The federal government has a duty to manage those lands in a manner that is in the public interest, not in the short-term interest of a particular industry. Is it wise to allow management of our public lands to be governed by a 19th century law that no longer reflects, and in fact, runs contrary to popular opinion? Sixty-seven percent of all Americans say no. [21]

To summarize, we recommend that Congress permanently end public lands giveaways to mining companies, impose a fair royalty for mining on public lands, create an abandoned mine cleanup program, and end the policy of giving mining companies first-use of our public lands. Thank you.

> "... After eight years in this office, I have come to the conclusion that the most important piece of unfinnished business on the nation's resource agenda is the complete replacment of the Mining Law of 1872."
>
> STEWART L. UDALL, SECRETARY OF THE INTERIOR, 15 JANUARY 1969

ENDNOTES

[1] Federal Land Management: The Mining Law of 1872 Needs Revision, GAO Report/ RCED-89-72, March 1989, 2.

[2] Multiple Conflicts Over Multiple Uses, Reform of the 1872 Mining Law: A Primer, John L. Dobra, ed. Terry L. Anderson, Political Economy Research Center, 1994, 39.

[3] The "Crown Jewel Mine Decision," U.S. Department of the Interior, Solicitor John Leshy, Mar. 25, 1999.

[4] Limitations on Patenting Millsites Under the Mining Law of 1872, U.S. Department of the Interior, Solicitor John Leshy, Nov. 7, 1997.

[5] Pub. L. No. 86-390, 74 Stat. 7 (1960), codified at 30 U.S.C. sec. 42 (b) (1994).

[6] The original bill was introduced in 1959 as S.2033. 105 Cong. Rec. 8734 (1959).

[7] American Mining Congress, The Mining Law and Public Lands, at 29, Jan. 11, 1968.

[8] Federal Land Management: The Mining Law of 1872 Needs Revision, GAO Report/ RCED-89-72, Mar. 1989, 3.

[9] Ibid

[10] Mineral Policy Center estimates. See MPC methodology in Mineral Policy Center, Burden of Gilt, June 1993, 28-31.

[11] Abandoned Hardrock and Noncoal Mines in the West, Draft, Western Governor's Association, March 24, 1998.

[12] Burden of Gilt, Mineral Policy Center, June 1993, 3.

[13] Dr. W. Thomas Goerold, Testimony before the Senate Subcommittee on Mineral Development and Production, Sept. 13, 1990.

[14] Ibid

[15] U.S. Environmental Protection Agency, Fish Kills Caused by Pollutioin, Fifteen Year Summary 1961-1975, Apr. 1978, 8.

[16] Gold mine cleanup progresses, The Denver Post, June 16, 1999.

[17] Summitville Consolidated Mining Company Inc./Galactic Resources Inc./Galactic Resources Ltd., Summitville Mine Site, State of Colorado, Division of Minerals and Geology, Feb. 1, 1993.

[18] Mining company sues state over cleanup plans, Great Falls Tribune, Sept. 3, 1998.

[19] American Mining Congress, The Mining Law and Public Lands, at 29, Jan. 11, 1968

[20] Mineral Policy Center, Thomas Michael Power, Not All That Glitters, 1993.

[21] Findings from a National Bipartisan Survey of Voter Attitudes Toward Mining Law Reform, Prepared for MPC et al by Frederic,/Schneiders Inc., September 1994.

BILLINGS GAZETTE

June 17, 2003

Stillwater purchase approved

Jan Falstad

DENVER—Given the choice between possible bankruptcy or being controlled by a Russian mining giant, Stillwater Mining Co. investors chose the Russians.

The sale of a majority of Stillwater stock to Norilsk Nickel of Moscow marks the largest purchase yet of an American corporation by a Russian company.

Following Monday's special shareholders meeting in Denver, Stillwater Chief Executive Frank McAllister said the stock sale will accomplish three things: "We brought financial stability to our company. We brought time to restore confidence in palladium and increase demand. And we preserved the jobs."

Stillwater vice president John Stark said 62 percent of the company's investors voted on the proposal. Investors representing 22 million shares voted "yes" to the Norilsk deal, while those holding about 5 million shares said "no."

Also Monday morning, McAllister announced the Federal Trade Commission had waived any antitrust concerns over the sale. With the federal roadblocks gone, McAllister said acquisition will be completed quickly.

"The closing will take place in the next two weeks," he said.

After the vote, union president Brad Shorey, reached in Columbus, said he was disappointed.

He said a majority of the 1,030 miners represented by Local 8-001 of the Paper, Allied Industrial, Chemical and Energy Workers International Union rejected the buyout.

He said he has never supported giving Norilsk controlling interest in the only American platinum and palladium mine and he worries about upcoming labor negotiations. The PACE contract expires in July 2004 and negotiations start soon.

"If we ended up in a labor dispute, what makes you think we're not doing exactly what they want?" Shorey said. "They could ride the line and play at good faith bargaining, while pushing us into a strike."

Norilsk Nickel has been facing labor problems at the mines it owns near the Arctic Circle.

Dave MacAusland, who owns a small business in Portland, Ore., and owns a cabin in Red Lodge, asked if investors could express their opinions before the final votes were cast Monday.

After a testy exchange, McAllister let MacAusland speak.

His emotional comments focused on why the United States would cede control of its only platinum and palladium mine—one of three major mines in the world—to any foreign corporation.

After speaking for 10 minutes, MacAusland received a round of applause from several other shareholders, the only applause of the meeting.

McAllister responded by saying the United States is already dependent on the world for many metals we formerly produced. America consumes half of the world's the world's palladium, but Stillwater produces only 5 percent. So, the U.S. is already a net importer of palladium, he said.

Another investor asked McAllister what alternatives to the Norilsk sale the company was considering.

"I cannot answer about alternatives," he said. "Obviously, we're in a very desperate and serious state at this time."

Stillwater's board twice rejected another offer by American investors to raise $100 million for Stillwater by selling more shares.

McAllister said that deal wouldn't have generated the cash needed to pay operating expenses and to pay down high debt.

As to fears the Russians would shut down the Nye or East Boulder mines in order to sell more of their own palladium, McAllister said that won't happen.

"They can't pick up the mines in the mountains of Montana and move them to Russia," he said.

No one representing Norilsk attended the Denver meeting at the Hyatt Regency Tech Center.

Carl Edstrom, a metals engineer and investor from Arvada, Colo., asked why the costs of production at Montana's mines jumped from $178 per ounce of palladium five years ago to $354 per ounce today.

McAllister blamed low production at the East Boulder mine, where the concentration of the ore was much lower than expected.

Platinum and palladium are interchangeable in the most common use for catalytic converters. So McAllister said he'll try to convince auto makers to switch from platinum, costing $680 an ounce, to palladium which has fallen from more than $1,000 an ounce to $180.

After the meeting, Dennis Schmidt, a retired wheat farmer and investor from Lawrence, Kan., said he still was doing a "slow burn" about the loss of value in his 20,000 shares.

"The management ought to be in jail for arranging this transaction and going through with it," he said.

McAllister said the company has enough cash to perhaps make it through this year, but has no "rainy day" fund left if something goes wrong.

This stock sale accomplishes a lot, which he said Montana politicians understand, but workers don't grasp yet.

"The employees are worried about their jobs just as I am" he said. "Well, it's done and they still have their jobs—the majority of them still have their jobs."

McAllister also said with the cash influx he's excited to pursue some other business activities outside of mining.

"Will we do something else? I hope so," he said.

He wouldn't comment on what ideas he has, except to say Stillwater will conduct palladium research and promotion and will expand a small sideline in Columbus that recycles catalytic converters.

McAllister is eligible for up to a $2 million transaction bonus for negotiating the Norilsk deal. The board will determine the exact amount. He also remains as Stillwater's top executive unless the new board controlled by Norilsk changes management.

JP Morgan, which handled the financing, gets $5 million.

The high costs of attorneys to handle Washington, D.C., legal and lobbying work haven't been tallied yet, McAllister said.

The powerful firm of Baker Botts lobbied the deal through the federal agencies. The lead lobbyist was James Baker, son of the former secretary of state.

Half of the $100 million in cash Stillwater will receive when the agreement is closed goes to pay off debt.

Norilsk is one of the world's largest metal producers and it mines palladium as a byproduct of copper and nickel production near the Arctic Circle.

Critics said mining palladium as a byproduct means Norilsk pays only $24 per ounce, a fraction of the $354 price tag at the Montana mines.

Even though Norilsk trades publicly in Russia, Russian secrecy laws don't allow the company to report how much palladium it is producing or has stockpiled.

Without that crucial data, critics say Norilsk's accounting isn't transparent or generally accepted.

When the deal was announced last November, Norilsk agreed to pay Stillwater $341 million for 51 percent of the company stock.

Stillwater would get $100 million in cash and the balance in metal, 877,000 ounces of palladium.

Since then, the price of palladium has dropped dramatically. That means the total sale price today is around $258 million or $83 million less than anticipated.

When the sale is completed, Stillwater will issue 45,463,222 new shares of common stock to a subsidiary of Norilsk called Norimet Limited.

Norilsk has the option of buying up even more shares up to 56 percent ownership if the stock price stays below $7.50 for two weeks after closing.

Stillwater's stock closed up 21 cents Monday at $4.83.

Union president Shorey said some miners who are financially able to quit said they will.

"They won't work for a Russian-controlled company," Shorey said. "At first I thought it was because they were older, but it's young and old people."

The mood at the mine Monday was watchful, Shorey said.

"From this day forward, we're going to be holding our breath to see what happens."

THE PLAIN DEALER

(Cleveland Plain Dealer)

May 22, 1994

FORTUNE HIDDEN UNDER DESERT

CLEVELAND FIRM WANTS TO BUY LAND WITH RARE ORE FOR $26,487

KEITH EPSTEIN PLAIN DEALER BUREAU

In the harsh desert of western Utah south of the Great Salt Lake, a desolate plateau of scrub and sagebrush stretches to the horizon. It is home to gophers, jackrabbits, rattlesnakes and the occasional coyote. Groundwater is too brackish for crops. The nearest house is 40 miles away.

"Nobody's ever wanted to homestead out there, and nobody ever goes to a place like that to live or have fun," observes industrial geologist Lee Davis. "To the ordinary fella, it would seem worthless."

But this desert, owned by America's taxpayers, conceals a vast fortune. Beneath 2,548 acres alone is a rare bertrandite ore, which, when processed, could be worth up to $15 billion.

Brush Wellman Inc., a Cleveland-based mining company, wants to buy those acres from the Interior Department's Bureau of Land Management for a mere $26,487—around $10.40 an acre.

That would be one of the richest lodes ever purchased from the government—worth potentially almost as much as last week's highly publicised sale of land with perhaps $18 billion in gold to a Canadian mining company. "The biggest gold heist since the days of Butch Cassidy," Interior Secretary Bruce Babbitt complained on Monday as he transferred 1950 acres in Nevada to Toronto-based American Barrick Resources Corp. for a mere $9,765. "It is a ripoff."

Babbitt is powerless to stop the sale because, for many years, Congress has declined to curb sales of precious public property. Dirt-cheap prices meant for grizzled miners of 122 years ago are still on the books.

Babbitt vows to insist upon replacing such sales with royalties that produce a "reasonable return" to taxpayers. But the House and Senate are far from agreeing on how to do so.

While Brush Wellman disputes some figures in this article, arguing that its high cost of exploration and development should be considered, the company agrees with the bottom line: By paying the government just thousands, it can obtain land worth millions.

"Nobody's disputing that," said Hugh D. Hanes, vice-president of environmental and government affairs. "It's a relatively small amount (returned to the Treasury) in comparison to what's in the ground."

Similar stories are unfolding elsewhere. In Oregon, for one, Denver-based Newmont Mining Corp. hopes to spend $1,560 for 60 acres containing most of an estimated $373 million in gold—and tens of millions more in silver.

For five months Western senators have been stalling final action to replace the 1872 mining law. While they dawdle, companies continue to wrest from public lands some $1.7 billion in gold and other minerals each year—with little or no direct compensation to the federal treasury.

That amount is only the General Accounting Office's guess. Government officials don't know how many millions of tons of precious minerals there are on federal lands, or what it's all worth—complicating any attempts to set a fair asking price.

"We don't have good facts," Interior Secretary Bruce Babbitt acknowledges. Babbitt told a congressional committee he was "really astonished" when trying to learn from his own employees the volume and value of mineral production on public lands, data that companies consider confidential.

"The answer I got was, 'we don't have any'," Babbitt said.

The tale of Brush Wellman's beryllium bonanza illustrates how government and corporations often cooperate to advance corporate interests, sometimes without fairly compensating taxpayers.

At stake is the future of 432 million acres of publicly owned lands—and the untallied treasures they contain. These lands are managed by the U.S. Forest Service and Bureau of Land Management.

This article was assembled using corporate documents, government records, interviews with company insiders, and a computer-assisted analysis of $131,336 in campaign contributions to members of Congress by Brush Wellman executives, lobbyists and committees. Checks are not illegal

The checks that flow each year to politicians from Brush Wellman and its employees are neither illegal nor random.

As Hanes acknowledged: "The company tends to support those people that are supportive of the company."

Two company supporters, Republican Reps. James Hansen of Utah and Paul Gillmor of Port Clinton, have received $41,400 from Brush since 1988.

An example of that support occurred last November, when the duo came within 45 votes of getting the company exempted from House mining rules requiring higher fees and royalties.

The congressmen argued that, without a financial break, the company might be forced to shut down the mines. America's only domestic source of beryllium would thus be jeopardized—and that, they said, elevated the exemption to a matter of national security.

But Pentagon documents show that the military is "over goal" in its stockpile of beryllium metal. Nor does it need beryl ore or a beryllium-copper alloy.

Far from being concerned about secure supplies of Brush Wellman's products, the Pentagon now wants to sell 24,221 tons of beryllium materials worth $122.9 million.

Explains Beth Offenbacker, spokeswoman for the National Defense Stockpile Center: "We feel we don't need it."

Brush Wellman, Gillmor and Hansen all say they intended only to fight for workers' jobs and the local economy.

They say they've done nothing that violates official rules or laws. By all accounts, they are correct—and for that they can thank not only campaign finance laws, but the General Mining Law of 1872, a statute dating to the administration of Ulysses S. Grant. Brush coverts bertrandite

Brush Wellman's sophisticated methods of extracting a fortune from the rocks by refining minerals into products with far-reaching technological applications would have dazzled even the hardiest, most crusty "Forty-niner."

Like other modern mining companies, Brush uses a chemical process to leach ore. Like other modern companies, it literally can move mountains and create cratered moonscapes.

Brush converts the bertrandite ore, found in Utah, Brazil, Africa, India and China, into light-weight, harder-than-steel beryllium. Its uses include satellites, nuclear reactors, airplanes, computers and cars.

Brush Wellman's products are in nuclear warheads. Every strategic missile in the U.S. arsenal contains beryllium, as did the "smart bombs" dropped on Iraq during the Gulf War.

What President Grant had in mind—around the time of Custer's last stand—was to encourage exploration and development of the western wilderness.

He wanted it used, opened up and settled. Thus, he offered to "patent" cheap land titles—for miners, not multinational conglomerates that sometimes scar the landscape and foul waters. But Utah and federal mining officials say that Brush Wellman's environmental record is impeccable—better than law currently requires.

For years, aggressive lobbying by mining corporations and railing by western politicians has thwarted attempts to revise the anachronistic law. Politicians such as Sen. Pete V. Domenici, R-N.M., often complain of a "war on the West."

But the Clinton administration's vow to seek market-based fees for use of federal lands for mining, grazing and timber-harvesting has created new pressure for reform.

"Just the first step in an assault on the west," Sen. Conrad Burns, R-Mont., complained during last year's debate on ranchers' fees.

Clinton, who made electoral gains in the usually Republican west, isn't likely to go too far in alienating his newfound western friends. Nor is the public lands brawl strictly east versus west.

Today's mines are more likely to be run from corporate boardrooms in Toronto or Cleveland. In fact, 33 companies now mining on public lands purchase $901 million in equipment and supplies that translates into jobs in the east, a fact not lost on eastern congressmen.

In Ohio, four companies besides Brush—Oglebay-Norton, Cleveland Cliffs, Dresser Industries and AEP—have mined on public lands. A fifth, Timken, is a major supplier of heavy mining equipment. House votes for royalty

In November, the House voted overwhelmingly to end the practice of selling cheap land to mining companies. Instead, companies would pay the government 8% of the processed minerals' value.

Amounting to some $100 million a year, such a royalty—significantly lower than the 12.5% coal, natural gas and oil companies have been paying for years—would hardly make a dent in the national debt.

And it would only begin to help pay for cleanups of thousands of old abandoned mines, many of which are fouling land and waters in locations throughout the Western United States.

Over time, such a royalty would have cost companies such as Brush Wellman hundreds of millions.

Thus, the industry supported a Senate-backed end to the land sales that imposes what environmentalists regard as a "sham" royalty—2% of the value of minerals before they are processed.

Companies also could deduct major business expenses.

Since then, key members of the House and Senate who are supposed to reach a compromise have only dawdled, and now lobbying has intensified. Last month, for instance, Hanes made his pitch to key members as part of an industry-sponsored "Hardrock Minerals Day."

Meanwhile, mining companies are wisely hedging their bets by seeking to "patent" their claims more quickly than ever. These are legal and administrative steps they must take before the government grants them title to the land.

As Congress began seriously debating an end to cheap land sales, mining companies scrambled to start applying for purchases. There's a "patent rush" out there—a frenzy to buy before it's too late.

In California and Nevada, for instance, more than 100 applications are pending; a few years ago, there were never more than 40 at any time.

"Companies know that something will eventually happen (in Congress), and with a foot in the door they may be able to keep that door open," observes Walter Phelps, who heads the Bureau of Land Management's office in Utah.

Interior Secretary Babbitt has tried to stall some sales—including the gold mine now lost to the Canadian company—until Congress passes a new mining law. A federal magistrate criticized Babbitt, saying this amouted to little more than a "shameful de-facto moratorium" on issuing mining patents.

Brush Wellman officers, meanwhile, hope that the company is far enough in the process that the eventual reform of mining laws will not apply; the company will be "grandfathered" in. If not, the company threatens a lawsuit.

"Our concern," says Hanes, "is protecting the investment we've made in the patenting process."

By that, he means the money spent on lawyers—about $1 million—to prepare eight different applications. Moreover, Hanes says the company has spent more than $8 million on exploration, and that investing in Utah was a "bet-your-company" move in the first place.

"Hardly a giveaway," he said. "We're not getting a free ride. The ore we mine on public lands requires a major up-front investment." Regula seeks reform

"A specious argument," responds Rep. Ralph Regula, R-Navarre. "First of all, they don't have to buy the land to mine it. And their investment is part of doing business whether they own the land or lease it. The fact is, nothing goes to the government."

For several years now, Regula has persuaded his House colleagues to ban the cheap land sales—only to be stymied in the Senate where, he complains, "westerners always kill any mining reform. It's outrageous."

To date, some 3.2 million acres of federal land—an area the size of Connecticut—have been sold, some for as little as a few dollars an acre.

And, Regula says, some land has even been sold back to the government after companies have squeezed what they could from the land—at a profit.

Davis, who was there on the plateau of sagebrush at the beginning of the beryllium mine, in 1968, complains that environmentalists and eastern politicians have distorted the issue.

"The idea that mining companies are getting ground cheap and not paying the government much is completely wrong," says Davis, who was Brush Wellman's chief geologist until retiring three years ago. "We pay an awful lot of state taxes and we help a lot of economies locally.

"And," he adds, "we'd be happy to give the land back to the government. After we've mined it."

MINERAL POLICY CENTER

PLAYING ON AMERICA'S NATIONAL SECURITY CONCERNS

MINING COMPANIES' FALSE ARGUMENT ON "STRATEGIC METALS AND MINERALS"

Seeking to boost profits under the guise of national security, mining corporations promote mining in the U.S. as a source of strategic minerals. In reality, most mining in the United States is for metals totally unrelated to national security. At the same time, metals mining puts U.S. taxpayers, western communities and our environment at risk.

- For years, the federal government has been selling off excess supplies of "strategic metals and minerals" to private entities.
- A storehouse of these materials was set aside after World War II under the Pentagon's strategic minerals program. But in 1992, Congress deemed most of the storehouse unnecessary and authorized the sale of the bulk of these materials.

National security should not be exploited as a distraction from the real issues and problems of hardrock mining.

- The vast majority of mining on U.S. public land is for nonstrategic purposes. 85% of gold mined in America is used for jewelry.
- Metals like gold and copper—which never appeared on the Pentagon's list of strategic minerals—generated more than $5 billion in 2001 for multinational mining companies.
- Many strategic metals and minerals must be imported, as they don't exist in America's ore bodies. No amount of mining on American land will make our country less dependent on other countries for these materials.

Hardrock mining operations endanger public health.

- Toxic chemicals released in mass quantities by mines include lead, arsenic and mercury.
- 335 million pounds of lead were released by mines in 2001, according to the industry's own reports to EPA. Lead poisoning can permanently damage a child's brain, nervous system and kidneys. It can impede growth and cause hearing loss, learning difficulties, vomiting, headaches and appetite loss.
- 336 million pounds of arsenic were released by mines in 2001, according to the industry's own reports to EPA. Arsenic consumed by humans can cause cancer of the bladder, liver and skin, according to the National Academy of Sciences. Arsenic is associated with birth defects, as well as damage to the human heart, blood vessels and nervous system.
- 4.3 million pounds of mercury were released by mines in 2001, according to the industry's own reports to EPA. Mercury is a potent neurotoxin. Children and infants exposed to mercury often experience delays in developing motor skills like walking and talking. New concerns center on increased numbers of women with elevated blood mercury levels, as this dangerous toxin can transfer through a placenta to a developing fetus, or to a newborn through breastfeeding.

American taxpayers are continually fleeced by multinational mining companies.

- Multinational mining companies have taken $245 billion worth of precious metals and minerals from land owned by U.S. taxpayers without paying a cent in royalties over the past 131 years. In contrast, coal, oil and gas companies pay 8% to 12.5% royalties—a total of $35 billion between 1994 and 2001 alone.
- A tax break amounting to $100 million per year is given to hardrock companies based on the declining value of minerals, even though the industry never pays for the minerals in the first place.
- Taxpayer costs to clean up more than 500,000 abandoned mines littering the U.S. landscape could total $32 billion to $72 billion.
- Taxpayers could be liable for $12 billion to clean up costs currently operating mines, according to the Center for Science in Public Participation.

Devastating environmental consequences of hardrock mining continue to grow.

- The nation's #1 toxic polluter is hardrock mining, releasing 2.8 billion pounds of chemicals like arsenic, lead, mercury and other heavy metals in 2001, according to the industry's own reports to EPA.
- Portions of 40% of western watersheds are contaminated by mining, according to EPA.
- Half a million abandoned mines litter the U.S., emitting acid mine drainage and other pollution.
- More than 70 hardrock mines already have become Superfund sites.
- Mine pollution includes everything from heavy metal contamination to cyanide spills and acid mine drainage. Streams seriously affected by acid mine drainage and heavy metal contamination are biologically dead.

Real reform of the 1872 Mining Law is needed to protect taxpayers and the environment.

- Strong environmental standards for hardrock mining should be enacted to protect western communities and valuable water resources.

- Hardrock mining companies should be required to pay an 8%-12% royalty to taxpayers for taking precious minerals from public lands—just like coal, oil and gas companies do.
- Giveaways of public land to multinational mining companies for less than $5 an acre should be permanently ended.
- Special places, like wilderness areas and sacred sites, should be protected from irresponsible and destructive mining practices.
- Western communities should be allowed to have greater input into mining proposals that will directly impact the economic and public health of their area.

○